U0891372

中国丹霞地貌及其研究进展

DANXIA GEOMORPHOLOGY OF CHINA AND ITS PROGRESS IN RESEARCH WORK

彭华 著

中山大學出版社
SUN YAT-SEN UNIVERSITY PRESS
·广州·

版权所有　翻印必究

图书在版编目（CIP）数据

中国丹霞地貌及其研究进展 = Danxia Geomorphology of China and its Progress in Research Work / 彭华著. —广州：中山大学出版社，2000. 8

ISBN 978 - 7 - 306 - 01702 - 4

Ⅰ. ①中…　Ⅱ. ①彭…　Ⅲ. ①地貌—丹霞类型—研究—中国　Ⅳ. ①P931

中国版本图书馆 CIP 数据核字（2000）第 65968 号

出 版 人：王天琪
责任编辑：李海东
封面设计：林绵华
责任校对：何　凡
责任技编：何雅涛
出版发行：中山大学出版社
电　　话：编辑部 020 - 84110771，84113349，84111997，84110779
　　　　　发行部 020 - 84111998，84111981，84111160
地　　址：广州市新港西路 135 号
邮　　编：510275　　传　　真：020 - 84036565
网　　址：http：//www. zsup. com. cn　E-mail：zdcbs@ mail. sysu. edu. cn
印 刷 者：佛山市浩文彩色印刷有限公司
规　　格：889mm × 1194mm　1/32　4. 5 印张　120 千字
版次印次：2000 年 8 月第 1 版　2019 年 6 月第 2 次印刷
定　　价：38. 00 元

如发现本书因印装质量影响阅读，请与出版社发行部联系调换

作者简介

彭华（1956. 1. 2—2018. 1. 8），安徽省砀山人，1982 年 1 月毕业于安徽师范大学地理系。中山大学地理科学与规划学院教授，博士生导师，享受国务院政府特殊津贴专家。主要从事丹霞地貌、旅游地理和旅游规划等教学与研究工作，独著、合著、主编出版著作 16 部，发表论文 150 余篇。

兼任国际地貌学家协会红层与丹霞地貌工作组主席，中国地理学会红层与丹霞研究工作组主任，国家住建部世界遗产专家委员会委员，中国地质学会旅游地学与地质公园分会委员，中国丹霞世界自然遗产专家委员会主任，丹霞山世界自然遗产、世界地质公园总工程师。

主持两项国家自然科学基金和一项国家科技基础性重点项目，是当代中国红层与丹霞地貌研究领域的学术带头人。主持召开了 12 届全国丹霞地貌旅游开发学术讨论会，主编会议论文集 11 部；策划、组织并主持了第一、二、三届丹霞地貌国际学术讨论会，是推动中国丹霞地貌研究国际化的代表性人物。

彭华 1999 年接任丹霞地貌旅游开发研究会理事长，重点推动了学科体系建设和丹霞地貌研究国际化工作。2000 年出版了丹霞地貌第一部中英文对照专著《中国丹霞地貌及其研究进展》，并启动了初步的国际交流。2003 年成功主持了丹霞山以“丹霞地貌类”入选首批世界地质公园，让丹霞地貌的名词（Danxia Landforms）首次登上 UNESCOG 和 IUGS 网站。2006 年以其专著内容为主体，丹霞地貌被以专章的分量纳入高校“十五”规划教材。策划和组织了 2009 年 5 月在丹霞山召开的首届“丹霞地貌国际学术讨论会”，2009 年 7 月“第七届国际地貌学大会”批准成立了“IAG 丹霞地貌工作组”，这两次会议标志着丹霞地貌走上国际最高学术舞台。2011 年 10 月和 2015 年 9 月，又先后在丹霞山和江郎山召开了第二届和第三届丹霞地貌国际学术讨论会。并应邀多次在国际地貌学会议上介绍丹霞地貌。同时，推动和主持了首个丹霞地貌国际合作研究项目和首个中美丹霞地貌国际对比研究，指导完成了第一个丹霞地貌国际对比研究的博士学位论文；协助指导了英国、美国和日本的研究生完成与丹霞地貌相关的1篇硕士学位论文和2篇博士学位论文。在国内外40多所高校及10多个科研机构、科学馆和博物馆等进行了60多场丹霞地貌和中国丹霞世界遗产科普讲座。

基于红层区存在着严重的地质灾害、土地退化和生态环境脆弱等问题，彭华近年来又开创性地提出了对红层问题开展“地下—地表—地上一体化研究”的新方向，使原先对丹霞地貌的景观研究，走上更加结合国计民生重大需求的研究道路，实现了学科面的拓展和质的提升。在其推动下，2011年7月，中国地理学会批准设立了“中国地理学会红层与丹霞研究工作组”；2013年8月，国际地貌学家协会批准原丹霞地貌工作组更名为“IAG红层与丹霞地貌工作组”。这标志着一个新学科被学界所认同。

彭华在丹霞地貌应用研究方面也取得显著成就。1987年以来，坚持以丹霞山为研究基地，致力于基础研究、应用研究与品牌培育，将丹霞山由一座地方名山培育成为世界名山。主持完成的《丹霞地貌风景资源可持续利用研究》系列成果获2006年教育部科技进步二等奖和2007年广东省科技进步二等奖。2006—2010年作为南方六省“中国丹霞”联合申报世界自然遗产专家组组长和首席专家，在组织协调、基础研究指导、申报材料和管理规划编制、提名地整治指导等方面发挥了核心作用，成功主持了申报研究工作，获六省联合申遗领导小组杰出贡献奖，广东省政府记一等功，江西、浙江等省政府特别贡献奖。主持完成的“丹霞山风景名胜区总体规划”获2013年度广东省优秀城乡规划设计“一等奖”和2014年度中国华夏建设科技“二等奖”。

PROFILE OF THE AUTHOR

Peng Hua (Jan 2, 1956—Jan 8, 2018), a native of Dangshan, Anhui Province, graduated from the Department of Geography of Anhui Normal University in January 1982. He was a professor and PhD supervisor of the School of Geography and Planning, Sun Yat-sen University, and an expert enjoying special subsidy from the State Council. He was mainly engaged in teaching and research on Danxia landform, tourism geography and tourism planning. He authored, co-authored and edited 16 books and published more than 150 papers.

He was also Chairman of the Red Beds and Danxia Geomorphology Working Group of the International Association of Geomorphologists (IAG), Director of the Red Beds and Danxia Geomorphology Working Group of the Geographical Society of China, member of the World Heritage Expert Committee of the Ministry of Housing and Urban-Rural Development, committee member of the Tourism Geoscience and Geopark Branch of the Geological Society of China, Director of the China Danxia World Natural Heritage Expert Committee, and chief engineer of Danxia Mountain World Natural Heritage and World Geopark.

He presided over two national natural science funds and one national science and technology basic key project, and was an academic leader in the research field of red beds and Danxia landform in contemporary China. He also presided over 12 national symposiums on Danxia landform tourism development and edited 11 conference proceedings. He planned, organized and presided over the first, second and third international symposiums on Danxia landform, and was a representative figure in promoting the internationalization of Danxia landform research in China.

Peng Hua took over as Chairman of Danxia Landform Tourism Development Research Association in 1999, focusing on promoting the construction of the discipline and the internationalization of Danxia Landform Research. In 2000, the first Chinese-English monograph on Danxia landform, *Danxia Geomorphology of China and its Progress in Research Work*, was published, and preliminary international exchanges were initiated. In 2003, he successfully presided over the selection of Danxia Mountain as the "Danxia Landforms" in the first batch of world geoparks, making the term "Danxia Landforms" on UNESCOG and IUGS websites for the first time. In 2006, Peng Hua's monograph on Danxia landform was adapted and incorporated as a special chapter into the "Tenth Five-Year Plan" teaching materials of colleges and universities. He planned and organized the first "Danxia Geomorphology International Symposium" held in Danxia Mountain in May 2009, and the "IAG Danxia Geomorphology Working Group" was approved and established by the 7th International Geomorphology Congress in July 2009. These two conferences marked the emergence of Danxia landform on the highest international academic stage. In October 2011 and September 2015, the second and third international symposiums on Danxia landform were held in Danxia Mountain and Jianglang Mountain respectively. He was also invited to introduce Danxia landform at international geomorphology conferences many times. At the same time, he promoted and presided over the first international cooperative research project on Danxia landform and the first international comparative research on Danxia landform between China and the United States, guiding the completion of the first doctoral dissertation on international comparative research on Danxia landform. He assisted and guided graduate students in Britain, the United States and Japan to complete one master's degree thesis and two doctoral degree theses related to Danxia landform. More than 60 popular science lectures on Danxia landform and China's Danxia World Heritage have been held in more than 40 universities and more than 10 scientific research institutions, science museums, and museums at home and abroad.

The red bed area is characterized with serious geological disasters, land degradation and fragile ecological environment. Thus Peng Hua has put forward a new direction of "underground-surface-and-above-ground integrated research" on the red bed problem in recent years. This method has enabled the original landscape research on Danxia landform to embark on a research road that is more in line with the major needs of the national economy and the people's livelihood, thus expanding the discipline and improving its research quality. Under its impetus, in July 2011, the Geographic Society of China approved the establishment of the "Red Bed and Danxia Research Working Group of the Geographic Society of China". In August 2013, the International Association of Geomorphologists (IAG) approved changing the name "Danxia Geomorphology Working Group" into "IAG Red Bed and Danxia Geomorphology Working Group". This marks the recognition of a new discipline by the academic community.

Peng Hua has also made remarkable achievements in the applied research of Danxia landform. Since 1987, with Danxia Mountain as the research base, Peng has devoted himself to basic research, applied research and brand cultivation, bringing world fame to this local famous mountain. The series of achievements of "Research on Sustainable Utilization of Danxia Landform Landscape Resources" he presided over and completed won the second prize for scientific and technological progress of the Ministry of Education in 2006 and the second prize for scientific and technological progress of Guangdong Province in 2007. From 2006 to 2010, as the group leader and chief expert of the "China Danxia" joint application group for world natural heritage in six southern provinces, he played a central role in organizing and coordinating group work, guiding basic research, preparing application materials and management plans, and guiding the remediation of nominated sites. He successfully presided over the application research work and won the

Outstanding Contribution Award of the leading group for the joint application of world natural heritage in six provinces, the first-class merit award of the Guangdong provincial government and special contribution awards from Jiangxi and Zhejiang provincial governments. The "Master Plan of Danxia Mountain Scenic Area" he presided over won the First Prize for excellent urban and rural planning and design in Guangdong Province in 2013 and the Second Prize for China's Huaxia Construction Technology in 2014.

在彭华同志先进事迹校园巡回图片展仪式上的讲话

（重印代序）

中山大学党委书记
陈春声

老师们、同学们：

大家下午好！

为深切缅怀彭华同志，号召全校师生向彭华同志学习，落实立德树人根本任务，推进一流大学建设，学校决定5月下旬至7月中旬在各校园举办彭华同志先进事迹校园巡回图片展。今天我们邀请彭华同志夫人丰秀荣老师、地理科学与规划学院师生以及学校相关职能部门负责人一起，举行简单庄严的仪式，作为此次巡回展览的开始。感谢大家参加。

彭华同志生前系我校地理科学与规划学院教授，是中国丹霞地貌科学研究的学科带头人，中国旅游地理和旅游规划领域的开拓者之一。他从1995年进入我校工作，为党的教育事业奋斗终生，体现了一名优秀共产党员坚守理想信念、无私忘我工作、勇于担当作为、努力追求卓越的高尚品质。他是我校爱国奋斗的时代先锋，为人师表的优秀典范，坚定文化自信的知识分子楷模。

本次展览分为不忘初心、献身丹霞，扎根大地、奋斗为民，学为人师、行为世范三个部分。生动刻画了彭华同志从知识青年下乡当农村生产队长，到丹霞山风景区当总工程师，再到中国丹霞地貌科学研究的学科带头人，他始终怀抱赤子之心、家国情怀，扎根大地、奋斗为民，用一生诠释了一名优秀共产党、一位优秀地理学家的道德真谛和崇高境界。彭华同志的一生是无私奉献给祖国教育事业的一生，他用实际行动诠释着为人师表的崇高风范。

习近平总书记曾指出："衡量一名共产党员、一名领导干部是否具有共产主义远大理想，是有客观标准的，那就要看他能否坚持全心全意为人民服务的根本宗旨，能否吃苦在前、享受在后，能否勤奋工作、廉洁奉公，能否为理想而奋不顾身去拼搏、去奋斗、去献出自己的全部精力乃至生命。"彭华同志正是以实际行动践行着党的根本宗旨和党员的客观标准，是践行弘扬社会主义核心价值观的杰出楷模，值得全校党员同志及师生员工学习敬仰。

老师们，习近平总书记先后对教师提出"三个牢固树立""四有好老师""四个引路人""四个相统一"的希望和要求，彭华同志正是这样的榜样。榜样的力量是无穷的，伟大时代呼唤伟大精神，崇高事业需要榜样引领。在实现中华民族伟大复兴的中国梦的伟大征程上，在新时期加快建设中国特色世界一流大学前进道路上，我们需要的正是彭华同志这样的榜样。正如罗俊校长在春季工作会议上强调的，希望我们所有的老师，向彭华同志学习，遵循党中央的希望和要求，一辈子做好教书育人这件事，做新时代用时间育人的好老师、用心思育人的优秀老师、用情怀育人的卓越老师。

今天仪式后，将在广州校区北校园、东校园，珠海校区进行巡回展出，各个二级党组织要积极组织党员进行观展学习。后期我们还将在全校范围内举办"弘扬爱国奋斗精神 建功立业新时代——彭华同志先进事迹报告会"，以期让更多师生学习彭华同志爱国奋斗、建功立业的精神，激发学校建设一流大学的精神动力。

斯人已逝，风范长存。老师们，同学们，全校上下特别是广大党员同志都应该向彭华同志看齐，学习他那种不忘初心、献身使命的忠诚担当，那种扎根大地、奋斗为民的家国情怀，那种学为人师、行为世范的崇高精神，坚定信念，团结一致，落实"四个服务"办学根本，落实立德树人根本任务，以时不我待、只争朝夕的精神，为新时期加快建设中国特色世界一流大学奋力前行，为实现中华民族伟大复兴的"中国梦"作出新的更大贡献！

2019 年 5 月 20 日

Speech at the Ceremony Commencing the Campus Roving Photo Exhibition of Comrade Peng Hua's Advanced Deeds

(Preface to the Reprinted Version)

Secretary of the Party Committee of
Sun Yat-sen University
Chen Chunsheng

Teachers and students,

Good afternoon, everyone!

We are here to cherish our deep memory of Comrade Peng Hua. In order to call on all the teachers and students of the university to learn from him, to implement the fundamental task of building people of high moral values, and to promote the construction of a first-class university, Sun Yat-sen University has decided to hold an itinerant photo exhibition of Comrade Peng Hua's advanced deeds on campus from late May to the middle of July. Today, we invite Comrade Peng Hua's wife, Feng Xiurong, teachers and students of the School of Geography and Planning, as well as the heads of the relevant functional departments of the university, to hold a simple and solemn ceremony as the beginning of this roving exhibition. Thank you for coming.

Comrade Peng Hua was a professor at the School of Geography and Planning of our university. He was the academic leader of Danxia geomorphology research in China and one of the pioneers in the fields of tourism geography and tourism planning in China. He entered our school in 1995 and worked for the education of the Party all his life, reflecting the noble quality of an outstanding Communist Party member who adheres to his ideals and beliefs, works wholeheartedly, has the courage to take on his deeds, and strives to pursue excellence. He is a patriotic vanguard of our school, an excellent model of teachers, and a model of intellectuals who hold firm cultural confidence.

The exhibition is divided into three parts. Part One is about how Peng Hua sticks to his original aspiration and devotes himself to the research of Danxia landform. Part Two tells about how Peng strives for the welfare of the people. Part Three is about how Peng works as a teacher and scholar. The exhibition vividly portrays the life story of Comrade Peng Hua: his experience as a rural production team leader while living in the countryside as an "intellectual youth", later as chief engineer of Danxia Mountain Scenic Spot, and then as the academic leader of Danxia geomorphological science research in China. He has always embraced a heart as pure as that of a newborn baby and noble feelings and emotions towards the country. He has worked for the people all his life, and has been an embodiment of the truthful morality and sublimity of an excellent Communist Party member and an excellent geographer. Comrade Peng Hua has selflessly dedicated his whole life to the education cause of the motherland, and he has shown us with practical actions what the lofty demeanor of a teacher is like.

General Secretary of the Central Committee of the Communist Party of China Xi Jinping once pointed out, "There are objective criteria to measure whether a Communist Party member or a leading cadre has the lofty ideal of communism. That is, it depends on whether he can adhere to the fundamental purpose of serving the people wholeheartedly, whether he can be the first to bear hardships and the last to enjoy comforts, whether he can work hard and perform his official duties honestly, and whether he can struggle regardless of his personal danger and give all his energy and even his life for his ideal." Comrade Peng Hua has been precisely carrying out the fundamental purpose of the Party and the objective criteria of a Party member with practical actions. He is an outstanding model for carrying forward the core values of socialism, and is worthy of the respect of all the Party members, other faculty members, and students in the whole university.

Fellow teachers, General Secretary Xi Jinping has successively put forward to teachers a series of requests: the "three firmly set-ups", the "four possesses", the "four good guides", and the "four unifies". Comrade Peng Hua is such an example. A fine example has boundless power. This great era calls for great spirit, and the noble cause needs the example to lead. On the great journey to realize the Chinese Dream of the great rejuvenation of the Chinese nation, and on the road of speeding up the construction of a world-class university with Chinese characteristics in the new era, what we need is examples like Comrade Peng Hua. As President Luo Jun stressed at the spring working meeting, all our teachers are expected to learn from Comrade Peng Hua, to follow the requirements of the CPC Central Committee, to do a good job in teaching and educating people all their lives, and to be good teachers who educate people with their time, minds, and feelings.

After today's ceremony, a tour show of the photos will be held in the northern campus and the eastern campus of Guangzhou, and the Zhuhai campus. Each secondary Party organization should actively organize Party members to observe and study at the exhibition. In the later stage, we will also hold a report on Comrade Peng Hua's advanced deeds with a theme of "Carrying forward Patriotism and the spirit to struggle and Making Achievements in a New Era" within the whole school. It is hoped that more teachers and students will learn from Comrade Peng Hua and will be patriotic and aggressive to make contributions so as to stimulate a spiritual motive force for the school to build a first-class university.

Though Comrade Peng Hua has left us, yet his demeanor is everlasting. Teachers and students! The whole school, especially all the party members and comrades, should learn from Comrade Peng Hua! We should learn from him his loyalty and responsibility of sticking to his original aspirations and dedicating himself to his mission; we should learn from him his love of the country and striving for the people; we should learn from him his learnedness and integrity. Let's strengthen our conviction, stick together, and implement the fundamental task of "four services" and building people with high morality. Let's seize every minute to strive to speed up the construction of a world-class university with Chinese characteristics in the new era and make new and greater contributions to the realization of the "Chinese Dream" of the great rejuvenation of the Chinese nation!

May 20, 2019

序 一

20 世纪 30 年代，地质学家陈国达教授（现中国科学院资深院士）在对广东仁化丹霞山及华南红层地貌作了大量研究之后，便以最具代表性的丹霞山命名了红层地貌的一个特殊类型——“丹霞地貌”。这一命名是岩石地貌学研究的一个预见，它虽然限指红色砂砾岩上发育的地貌类型，但却同时突出了它在岩石地貌学研究中的地位，唤起了地貌学家对它更多的注意。事实也确是如此，从那以后，尤其是 80 年代以来，丹霞地貌研究引起了地学界的广泛关注，也引起了旅游管理部门和有关风景区的很大兴趣。90 年代初，他们发起成立的“丹霞地貌旅游开发研究会”，可以说是一个产研结合的学术、科普和开发性群众组织，这样的由不同学科的学者、政府官员、风景区管理人员甚至旅游企业家组成的研究组织在中国是唯一的。它也是一个某种形式上把科研成果转变成生产力的融合组织，为中国地貌学的发展走出了一条新路——以开发养科研、促科研。因此，它具有强大的生命力。其中代表人物是中山大学已 73 岁高龄的黄进教授，他的足迹已踏遍了中国的东西南北，考察了 450 余处丹霞地貌，他对丹霞地貌研究事业的执著态度和奉献精神令人钦佩和敬仰，他的事迹不仅在地学界而且在很广泛的范围为人所知。更令人欣慰的是，他培养了一位富有开拓性的接班人，即本书的作者彭华教授。他们卓有成效的工作加上各地学者的努力，研究会成立以来的 8 年时间里，召开了 6 届讨论会，出版了 6 本论文集，发表了近 200 篇论文，这样一种发展形势在国内是超前的，在国际上也是少有的。我为他们的成就而高兴，为能在这里说几句话而自豪。所以，尽管我不善于作序，但对彭华教授的要求还是欣然应允。

从地貌学分类上看，一直有“构造地貌”和“岩石地貌”两大类的划分。岩石地貌学中石灰岩、砂岩、砂砾岩、花岗岩、流纹岩、玄武岩甚至黄土皆有各自特有的属性和独特的地貌造型。在许多地方，以地貌为主体的风景资源都已经或正在成为重要的旅游景点或景区。而丹霞地貌不但造型奇特——这也许是各类岩石地貌共有的——而且它鲜明的红色是其他各类地貌所没有的，也许这正是它“鹤立鸡群”的原因，受到人们普遍地关注。在国外如美国的犹他州、阿利桑那州也有类似的地貌，但没有特别的命名。红层和丹霞地貌在中国分布广泛，据黄进统计已超过 500 处，它是在中国土生土长的地貌类型，而这种分类不论是从科学上还是从应用上，都是十分必要的。所以中国成为“丹霞地貌”的诞生地是很自然的。

彭华等人是力图将丹霞地貌推向世界的年轻一代，我想这也应该是老一代地貌学家的愿望。这本小册子系统而简洁地介绍了中国丹霞地貌研究进展，我希望它能给国际同行们一个比较简明而完整的信息，告诉人家我们做了什么，这一步正是以往所缺乏的。但愿我们的工作能够获得国际同行的认可，争取更多的交流，以促进岩石地貌学和旅游事业的发展，并希望在不久的将来，能够在我们的国家召开丹霞地貌国际学术讨论会。

崔之久

（中国地理学会地貌与第四纪专业委员会主任）

2000 年 4 月 21 日 于北大燕园

PREFACE 1

In the 1930s, after an extensive and intensive study of the geomorphology manifested in Mt. Danxia, Renhua, Guangdong and in the red beds of South China, Professor Chen Guoda, the geologist and academician of The Chinese Academy of Sciences, named a special geomorphologic type developed in red bed as Danxia Landform after Mt. Danxia, the most typical place where such landform occurs. This nomenclature is a foresight in the study of petrographic geomorphology. Although it refers only to the geomorphologic type developed in red sandstone and conglomerate, yet it also highlights its position in the research of petrographic geomorphology, thus attracting the attention of many geomorphologists. This is truly what things are.

From then on, and in particular since the 1980s, the research on Danxia Geomorphology has been arousing considerable interest both throughout the earth science circle, and among the tourism administration agencies as well as the scenic spots concerned. The Danxia Landform and Tourism Development Research Society initiated and set up by them can be said as an academic, scientific popularization, exploration mass organization combining production with scientific research. Such a research organization composed of scientists of various disciplines, government officials, managerial personnel of scenic spots, and even tourism enterprisers is unique in China. In a certain form, it is also an organization turning the scientific research results into productive forces, thus breaking a new path for the development of geomorphology: to finance and promote scientific research through the development of tourism resources. Therefore it is full of vitality and has a bright future. The typical representative of this organization is Professor Huang Jin of Zhongshan University who is now of the advanced age of 73. His track has covered all the four corners of China to investigate more than 450 Danxia Geomorphologic sites. His spirit of inflexible approach and constant contribution to geomorphologic research is admirable and respectable. His achievements are well known not only in the earth science circle but also well beyond it. It is gratifying to know that he has trained a successor full of pioneering spirit, i.e. the writer of this book, Professor Peng Hua. With their fruitful work and the efforts of researchers from the various places, and in the past 8 years since its establishment, the Research Society has convened 6 symposia, and published 6 volumes of proceedings and nearly 200 papers. Such a development is fast in China and rare in the world. I am pleased about their achievements, and proud to be able to put in a few words here. Although I am not good at writing a preface, yet I am glad to do so in response to Professor Peng Hua's invitation.

In geomorphologic classification, there have always been two broad divisions namely the "structure geomorphology" and the "petrographic geomorphology". In petrographic geomorphology, limestone, sandstone, sandy conglomerate, granite, rhyolite, basalt and even loess all have their own properties and geomorphologic forms. Many places have become scenic spots or resorts for tourists due to their landscape resources, especially the geomorphologic resources. The peculiar appearance of Danxia Landform may be shared by other kinds of rock, while its fresh red color is, however, unique. This is perhaps why Danxia Landform can stand out among others, and attracts much attention. Similar landform is present also in other country, e.g. in Utah and Arizona of the United States, but they have not given it a special name. Red bed and Danxia Landform are widely distributed in China. According to Huang Jin, over 540 locations of such landform have been discovered. It is a geomorphologic type born and brought up in China. This classification is necessary no matter in scientific or in application point of view. Therefore it is only natural that China is the birthplace of Danxia Geomorphology.

Peng Hua and others are a young generation endeavoring to introduce the Danxia Geomorphology into the world, which, I presume is also the hope of the geomorphologists of the old generation. This booklet deals systematically and concisely with the progress in China in the research of Danxia Geomorphology. I hope it will give our international colleagues relatively brief and complete information about what we have done. This is exactly what was lacking in the past. If only our work can be accepted by our international colleagues, and we can have more exchanges with them, in order to accelerate the development of petrographic geomorphology and tourism . We hope in the near future an international symposium on Danxia geomorphology can be held in China.

Cui Zhijiu
(Chairman of Commission on Geomorphology, Geographical Society of China)
in Yanyuan of Beijing University
2000/4/21

序 二

我关注丹霞地貌始于研究区域地质。记得1940年读李希霍芬(F. von Richthofen)之 *China*，知他认为粤北曾为红层被覆。之后学者们更分为丹霞层和南雄层等。但余等调查后发现各红层盆地分别发育在古盆地中，山间盆地多为砂砾岩，河谷多为泥砂岩。于是1945年写《南雄、丹霞层位新见解》一文，认为两地层可能是“同期异相”之物。毕业论文即以仁化县丹霞地形为题，称之为“红石公园”，想来距今已近60年矣。虽30年代后期，陈国达先生曾提出“丹霞地形”之名词，但那时并无人将其作为一种独立之类型进行研究，大都是从地质学或地层学角度，讨论红层的地貌表现及地貌发育之差异。丹霞地貌一词也没有成为通用之学术名词，包括本人在内，当时的研究也只是皮毛而已。

50年代始，我把丹霞地貌作为一种独立之地貌类型写进有关文章，并在之后的著作中，将丹霞地貌归类于岩石地貌之红层地貌类，进行了专门论述。大概从那时起，丹霞地貌之概念为国内更多的地学工作者所接受。遗憾的是，吾等不专，对丹霞地貌没有再进行深入研究。

令余欣慰者，80年代以来，黄进教授不懈进取，专攻丹霞，历尽艰辛，踏遍祖国山水，建树甚丰。90年代初黄进兄又倡导、组织全国学者，成立丹霞地貌旅游开发研究会，结合旅游开发，工作扎实而有生气，使丹霞地貌学科之研究得以空前发展。更喜事业后继有人，今读彭华教授新作，知当前丹霞地貌研究之进展，颇有井蛙之感，亦知其力图开拓之气魄，冲出国门之雄心，吾等不胜欣喜。余知美国犹他州天生桥区（Arches）、法国撒克逊区、澳洲等地均有丹霞地貌分布，而我国之多之美，全球无比。将丹霞地貌推向世界，与国际研究接轨，十分必要且条件已经成熟。

丹霞地貌是一种难得之旅游资源，国外的丹霞地貌区也多是著名风景旅游区，中国的丹霞地貌亦必将为国家之旅游事业发挥巨大作用，旅游部门更应重视之。余知彭兄又是旅游开发方面的专家，亲自主持了丹霞山的开发建设，相信丹霞地貌研究会在他的主持下，必将在基础研究和应用研究之结合中，为丹霞地貌学科之发展大展宏图。吾等年事已高，难以再有作为，谨祝他们为国家建设和旅游发展作出更大贡献，他们的成果更加充实地貌科学体系，早日获得国际同行之认可，使丹霞地貌这个在中国土生土长的地貌类型早日走向世界。

本书系统总结和介绍了国内丹霞地貌研究成果，必将在同行中产生巨大影响，其功大矣。故乐为序。

曾昭璇

2000年4月23日

PREFACE 2

My concern with Danxia Geomorphology began with a research on regional geology. I remember when I read F. Von Richthofen's *China* in 1940, I understood he held that northern Guangdong had been covered by red beds. Subsequently some research workers divided further the red beds into Danxia Bed and Nanxiong Bed. After an investigation, however, we noted that the various red basins were developed separately in various old basins, while intermontane basins were mostly filled by sandy conglomerate and valleys, by pelitic sandstone. Thus, I wrote a paper *New views on Nanxiong and Danxia Beds* in 1945, suggesting that these two Beds were likely to be equivalent although heteropic in facies. In my graduation thesis entitled "Danxia landform in Renhua County", I called it Red Stone Park. This, I think, happened nearly 60 years ago. In the late 1930s, although Mr. Chen Guoda proposed the term Danxia Landform, no one studied it as an independent type at that time, except for some discussions made on the geomorphologic manifestations and the difference in geomorphologic development in red beds mostly from a geological or a stratigraphical angle. Danxia Landform did not become a popular term either. All the research work on Danxia Landform at that time, including mine, was superficial only.

Beginning in the 1950s, I wrote in as an independent type the Danxia Geomorphology to a relevant paper, and in my subsequent works, dealt specially with the Danxia Landform as the red bed class in rock geomorphologic classification. Approximately from then on, the concept of Danxia Geomorphology have been more widely accepted by workers of earth sciences in China. To my regret, I have not made probing study into this subject any longer since then.

What gratifies me, however, is that since the 1980s, Professor Huang Jin have spared no effort to keep forging ahead in studying specially the Danxia Landform. He has experienced many hardships, traversed the length and breadth of the land and scored many achievements. In the early 1991s, Huang Jin organized relevant research workers in China to set up The Danxia Landform and Tourism Development Research Society to go parallel with tourism development. His work has been carried out sturdily and full of vitality, so that the research on Danxia Geomorphology can soar to unprecedented height. What gratifying more is that there are worthy successors to the cause. By reading Professor Peng Hua's new works, I know about the progress on Danxia geomorphologic research, and I feel myself in the past rather liked a frog at the bottom of a well able to see only the little patch of sky above. I also very glad to know he has the pioneering spirit and ambition to introduce this type of geomorphology into the world. To my knowledge, Danxia Landform is recognizable also in the Arches area of Utah, U.S.A., the Saxon area, France, and Australia, although this type of geomorphology in China is incomparable as far as its wide distribution and beautifulness are concerned. To introduce it into the world and to link it up with international research activities are necessary and the opportunity for doing so is now ripe.

Danxia Landform is a rare tourism resource, since Danxia Geomorphologic areas in foreign countries have mostly become famous scenic resorts. Therefore, Danxia Landform in China will be bound to play a big role in China's tourism, and tourism department should attach more importance to it. I understand Peng Hua is also a specialist in tourism development, who has personally taken charge of the development and construction of Mt. Danxia. I believe under his management, the Danxia Geomorphologic Research Society will ride on the crest of success in the development of the discipline of Danxia Geomorphology in the context of a combination between basic study and application research. As I am advanced in age, I am unable to do something worthwhile. I only hope they will make a greater contribution to China's construction and tourism development, their results will further enrich the scientific system of geomorphology and will be accepted earlier by our international colleagues, and Danxia Geomorphology, a special type born and brought up in China, will be introduced into the world soon.

This book summarizes systematically the research results on Danxia Geomorphology in China. It will certainly produce tremendous effect among the colleagues. As the achievement is great, I am glad to write this preface for it.

Zeng Zhaoxuan

April 23, 2000

序 三

1928 年，冯景兰对粤北丹霞山由（丹霞层）红色砂砾岩所形成的奇险壮观的丹崖赤壁地貌作了生动的描述，并命名了下第三系丹霞层。1939 年，陈国达把与丹霞山（性质）类型相同的地貌命名为“丹霞地形”。以后，吴尚时、曾昭璇对粤北等地的丹霞地貌作了进一步的研究和宣传，使丹霞地貌成为被广为接受的地貌类型。

作为一个后辈，我认为这种由中国学者发现并命名的丹霞地貌，应由中国学者继续作全面系统的深入研究并推向世界。数十年来，我对中国的丹霞地貌不断地进行调查研究，迄今为止，对中国已发现的 540 多处丹霞地貌的 450 余处进行了现场考察，对中国丹霞地貌形成发育的许多方面进行了探究。现正在撰写《中国丹霞地貌》一书，希望此书出版后，能为中国丹霞地貌走向世界尽我的一份绵薄之力。

彭华同志和我一样，多年来有志于把中国的丹霞地貌推向世界。他为了宣传中国丹霞地貌及研究进展，不辞辛劳，吸取了中国许多学者研究丹霞地貌的主要成果，加上他自己多年对丹霞地貌的研究心得，写成了这本《中国丹霞地貌及其研究进展》，反映了中国当前丹霞地貌的研究水平。此书简明扼要，第一次系统介绍中国丹霞地貌及其研究进展，是中国丹霞地貌推向世界的开端，是可喜可贺的事！

真正把中国丹霞地貌推向世界，恐怕还要经过几代人的努力，我与彭华同志愿为之奋斗！特以此为序。

黄　进

2000 年 5 月 17 日

PREFACE 3

In 1928, Feng Jinglan made a vivid description on the precipitous and magnificent red cliff and red wall geomorphology formed in red sandy conglomerate stratum in northern Guangdong, and termed this stratum as Lower Tertiary Danxia Bed. In 1939, Chen Guoda named the geomorphology similar to Mt. Danxia type as Danxia Landform. Subsequently, Wu Shangshi and Zeng Zhaoxuan conducted further studies and propaganda on the Danxia Landform of northern Guangdong, thus this geomorphologic type have since been widely accepted.

As a junior, I believe this Danxia Geomorphology discovered and named by Chinese scientists should be studied systematically, thoroughly and continuously also by Chinese research workers and introduced into the world. In the past several decades, I have constantly carrying out investigations and studies on China's Danxia Geomorphology. To this day, I have already carried out on-the-spot investigation in more than 450 out of the 540 Danxia Landform sites found so far in China, and have been probing into many aspects in regard to the formation and evolution of China's Danxia Geomorphologic type. I am now writing a book entitled *Danxia Geomrophology of China*, which, I hope will aid in introducing the Danxia Geomorphology into the world after its publication.

Peng Hua, like me, has had the ambition for many years to introduce China's Danxia geomorphology into the world. With a view to publicize the Danxia geomorphology of China and the progress in research work, he has taken the trouble to assimilate the chief results of Danxia Geomorphologic research from may Chinese workers. On this basis and also based on his own research results accumulated for many years, he has written this book entitled *Danxia Geomorphology of China and its progress in research work,* which mirrors the current research level in China concerning this geomorphologic type. This book is concise and to the point. It deals systematically for the first time with China's Danxia Geomorphology and the progress in research work, and marks the beginning of introducing Danxia Geomorphology into the world. It is a matter for congratulation!

I think it needs the efforts of several generations to actually introduce China's Danxia Geomorphology into the world. Both Peng Hua and me are willing to strive for it! This will serve specially as a preface.

Huang Jin

May 17, 2000

前 言

目前地貌学教科书上一般不专门讲丹霞地貌，只是上课的时候听老师说过，但不知其所以然。记得大四的时候我的同学朱诚拿了一张齐云山的地形图，说是丹霞地貌，想写文章，要我帮他作出立体图。因为没有见过丹霞地貌，所以作出的图只有起伏的立体效果而没有丹霞地貌的味，令这位老兄大失所望。

我第一次认识丹霞地貌是 1981 年底在安徽师范大学地理系作毕业论文的时候。我非常感谢导师卢村禾教授和闵煜铭教授给了我在皖南考察的机会，使我有幸在齐云山上住了一个星期。但这次并不是进行地貌研究，而是素描写生。齐云山是一处发育良好的丹霞地貌，是中国四大道教胜地之一。我被它优美的自然风光和丰富的文化遗迹所吸引，因此而改变了行程计划，在山上多住了 3 天。后来我根据地形图定位拼绘了齐云山的全景立体图，这才使得朱诚连声叫好。没想到，我画的第一幅名山立体图竟是从齐云山开始，从丹霞地貌开始。后来我为齐云山画过几个版本的导游图，并分别被收入不同的书籍和《齐云山志》。也因此与丹霞地貌结下了缘。

但研究丹霞地貌还是从 1987 年开始，我跟着陈传康先生做粤北的地质地貌及旅游开发研究，丹霞山、金鸡岭等丹霞地貌风光是其旅游资源的重头戏；1988 年有幸与黄进教授相识，后来又陪他再上齐云山考察。两位先生的言传身教，才使我真正走上丹霞地貌与旅游开发的研究道路。后来我在丹霞山做过较多的工作，尤其是 1992~1995 年我在丹霞山做了 3 年多的总工程师，有机会更多地了解丹霞山和丹霞地貌的有关信息，也有机会为丹霞地貌研究事业多做一些工作。

从 1991 年在黄进教授和陈传康教授的组织下成立全国丹霞地貌旅游开发研究会以来，一大批地学工作者投身其中，我国的丹霞地貌研究事业开始了一个新的发展时期。特别是近 20 年来，黄进教授为中国的丹霞地貌研究事业跑遍了全国各地，考察了 450 多处丹霞地貌。他一丝不苟的治学态度、坚韧不拔的献身精神、实事求是的工作作风对同行们产生了巨大的鼓舞和鞭策。

在前人研究的基础上，国内学者通过近十几年的努力，丹霞地貌科学体系已日趋成熟，社会影响也逐步扩大。但是因为与国外的交流甚少，目前国际上还没有将丹霞地貌作为一个独立的地貌类型进行研究。因此，争取与国外同行互通信息显得十分必要。如何使丹霞地貌形成更广泛的国际影响，确立中国土生土长的丹霞地貌在国际地貌学分类系统中的地位，成为我们义不容辞的任务。

本来黄进教授正在撰写《中国丹霞地貌》一书，故我一直没有写此书的计划。但是为了不错过在 2000 年南京国际地貌学专题会议上宣传中国丹霞地貌的机会，我向大会筹备组提出了进行丹霞地貌研究成果展和组织丹霞地貌考察线路的申请，得到了大力的支持。但是怎么推出，拿什么推出成了难题。1999 年第六届全国丹霞地貌旅游开发学术讨论会上，崔之久教授提出要有一个简单而又系统的、图文并茂的中国丹霞地貌及研究状况介绍。在黄进教授的支持与指导下，我开始准备这一工作。由于时间仓促，系统总结几十年的研究成果是十分困难的，这本小册子只能概略地介绍有关方面的主要认识和主要成果，我希望它能让国外同行对中国丹霞地貌及研究进展有一个初步的了解，也期盼着同行的回应、交流与支持。

感谢黄进教授的支持与指导，感谢曾昭璇教授、黄进教授和崔之久教授在百忙之中为本书作序，感谢中山大学出版社尤其是李海东先生的大力支持并及时编辑出版，感谢地质出版社的蔡卫东先生曾经给予的帮助，感谢丹霞山风景名胜区管理委员会资助出版，感谢所有为本书出版而付出劳动和给予支持的人，感谢所有关心与支持丹霞地貌研究事业的人！

彭　华

2000 年 7 月 20 日　于中山大学

FOREWORD

As a rule, the current textbook of geomorphology does not specially involve Danxia Geomorphology, which has only been talk about by the teacher in class, so the students generally do not know it in detail. I remember when I was the fourth-year student, my fellow-student Zhu Cheng showed me a Qiyunshan Topographic Map, saying that it was for Danxia Landform. He wanted to write a paper and asked me to draw a space diagram for him. Because I had not seen any Danxia Landform before, so the diagram I drew for him had only the relief effect without any taste of a Danxia Landform. This greatly disappointed him.

The first time I know about Danxia Landform was at the end of 1981, when I was writing my graduation thesis in the Geographic Department of Anhui Normal University. I was very grateful to my teachers Prof. Lu Cunhe and Prof. Min Yuming for giving me the opportunity to investigate in southern Anhui, so I could be fortunate enough to stay a week in Qiyunshan. But this time I only took sketch from nature instead of studying geomorphology. Qiyunshan is a site where Danxia Landform is well-developed, besides it is one of the four major Taoist resorts in China. Fascinated by its excellent natural scene and rich cultural remains, I changed my plan and stayed three days longer on the mountain there. Later, based on a topographic map, I drew a space diagram of the overall view of Qiyunshan, for which Zhu Cheng rained praises on me. I hardly expected the first famous mountain three-dimensional diagram I drew was actually for Qiyunshan and for Danxia Landform! Subsequently, I drew several editions of tourist maps for Qiyunshan, which was collected into different books including the *Qiyunshan Chronicles*. Therefore I became tightly bound to Danxia Geomorphology.

However, I did not actually began to study Danxia Geomorphology until 1987, when I accompanied Mr. Chen Chuankang to carry out a research program on geology, geomorphology and tourism development in northern Guangdong. The Danxia Geomorphologic scenes in Mt. Danxia and Jinjiling are the main tourism resources. In 1988, I was fortunate to make acquaintance with Prof. Huang Jin, and to accompany him later to go up Qiyunshan again on an investigation tour. Only I learned by precept and example from these two masters, could I really begin my career of research on Danxia Geomorphology and tourism development. Afterward, I worked rather a lot in Mt. Danxia, and in particular in 1992-1995, I was the Chief Engineer of Mt. Danxia for over three years. Hence, I had the chance to understand more of Mt. Danxia and to get access to much more information concerned with Danxia Geomorphology., and had also the chance to do more work for the Danxia Geomorphologic research career.

Since 1991 when The Danxia Landform and Tourism Development Research Society of China was established under the organization of Messrs. Huang Jin and Chen Chuankang, a mass of earth science workers have joined in it, while Danxia Geomorphologic research entered a new development period. Especially in the past 20 years, for making contribution to the cause of Danxia Geomorphologic research, Prof. Huang Jin have traveled all over China to investigate nearly 450 Danxia Geomorphologic sites. His meticulous scholarship, persistent devotion and practical and realistic style of work have greatly inspired and encouraged his colleagues.

Based on the achievements of the forerunners, and through the research work of Chinese scientists in the past decades, the scientific system of Danxia Geomorphology is getting ripe with every passing day, and its social influence becomes greater and greater. However, as not much international exchange has been made, the research on Danxia Geomorphology has not yet begun in foreign countries. Therefore, inter-communications with foreign colleagues are highly necessary. To widen Danxia Geomorphology's international influence, and to establish a position in international geomorphologic classification system for Danxia Geomorphologic type, which is born and brought up in China, are our bounden duty.

Since Prof. Huang Jin was writing a book *Danxia Geomor-phology of China*, I did not originally have any plan of writing this book. However, in order not to miss the chance to populize China's Danxia Geomorphology during the 2000 Nanjing International Geomorphologic Symposium, I submitted a proposal to the preparatory group to organize an exhibition on the research results and a field trip in relation to Danxia

Geomorphology, which immediately won wide support in response. But the difficult problem was how and what had we in hand to organize it? In the '99 6th All-China Symposium on Danxia Geomorphology and Tourism Development, Prof. Cui Jiuzhi suggested that we must have a booklet with excellent pictures and assay to deal systematically and concisely with China's Danxia Geomorphology and the progress in research work. Backed up and guided by Prof. Huang Jin, I accepted the task to prepare this booklet. As time presses, it is very difficult to summarize systematically several decades' research results. This booklet can only deal briefly with the relevant major results and our major knowledge of Danxia Geomorphology. I hope it will help our foreign colleagues to understand preliminarily the Danxia Geomorphology of China and the progress in research work, and I also hope it will meet with responses from colleagues. Exchange and support are in particular welcome!

I am grateful to Prof. Huang Jin for his support and guidance, to Mr. Zeng Zhaoxuan, Huang Jin and Cui Jizhi for their writing of the prefaces, to the Zhongshan University Publishing House, especially to Mr. Li Haidong for his energetic support in time, furthermore, to Geological Publishing House, especially Mr. Cai Weidong who has given me help greatly, to Mt. Danxia National Park Administration committee who has imbursed the book's publishing. Thanks to all those who have supported and aided in the publishing of this booklet; thanks to all those who show solicitude for and support the Danxia Geomorphologic research career!

Peng Hua

In Zhongshan University 2000/7/20

目　　录

Contents

1 红色的石头，红色的山崖，地质学家在这里命名了一种特殊的地貌——丹霞地貌

Red rocks and red cliffs, Geologists have named a peculiar landform occurring here as Danxia Landform

在中国广东省韶关市东北，有一片神奇的山地，红色的石头，红色的山崖，看去似赤城层层、云霞片片，古人取其“色如渥丹、灿若明霞”之意，称之为丹霞山。

丹霞山位于南岭山脉南侧的一个盆地中，红色山群约 280 km^2，呈峰林结构，高下参差，错落有致，有大小石峰、石堡、石墙、石柱、石桥等奇峰异石 600 多座，主峰巴寨海拔 618 m。整个山区宛如一方红石雕塑园，故称“中国红石公园”。 如今这里是国家重点风景名胜区和国家地质地貌自然保护区。

很早以前，人们就已经发现丹霞山与众不同。20 世纪 30 年代，中国地质学家、现中国科学院院士陈国达教授在对丹霞山及华南地区的红层山地作了深入研究之后，以发育最典型的丹霞山为名，将这一类地貌命名为“丹霞地貌”，并逐渐被国内学术界广为接受。经过几代人的研究，在中国已经发现了 540 多处丹霞地貌。丹霞地貌研究的理论与实践意义，日益为人们所认识，尤其是其特殊的风景资源价值，得到了社会各界的广泛关注。

In the northeast of Shaoguan City of Guangdong Province, China, there is a wonderful hilly terrain where red rocks and red cliffs are distributed here and there, appearing like many red walls and rosy clouds. The ancients called it Mt. Danxia, because this place is as deep red as "dan" and as red bright as "xia" (in Chinese, "dan" and "xia" mean "red pills" and "rosy clouds" respectively).

Mt. Danxia is situated in a basin on the southern side of the Nanling Mountains, with its hills covering an area of some 280 km^2, displaying a hoodoo or needle karst structure. More than 600 stony peaks, stony fortresses, stony walls, stony pillars and stony bridges of varied sizes and heights are well proportioned. As the entire hilly area looks like a red stone sculpture park, it is known as The Red Stone Park of China, and has now become a national park and a geological and geomorphologic natural protection area of China.

People have long ago known that Mt. Danxia is out of the ordinary. In the 1930s, following an intensive study on Mt. Danxia and other red bed terrains in South China, a Chinese geologist Prof. Chen Guoda, now an academician of the Chinese Academy of Sciences, termed that kind of landform as Danxia Landform after Mt. Danxia where such landform developed most typically. This term has gradually been widely accepted by the Chinese academic circles. Through the studies of several generations, Danxia Landform has been discovered in more than 500 places in China. The theoretical and practical significances in the study of Danxia Landform have gradually been brought to light. Its value as a special scenic resource in particular has attracted the attention of all walks of life in the society.

2 初创，成型，发展：丹霞地貌学在中国的土地上成长

Initiation, taking shape, and development: Growing up of Danxia Geomorphology in the land of China

1928 年，中国地质学家冯景兰在广东北部考察时发现，构成丹霞山的红色砂砾岩与周围的泥质红层发育的地貌差异显著，故将构成丹霞山的红层命名为“丹霞层”。1938 年陈国达研究了丹霞山的红层和地形发育特征，首次提出“丹霞山地形”的概念。1939 年陈先生正式使用“丹霞地形”的术语，以后丹霞层、丹霞地形的概念便被沿用下来。此后，不少学者对中国南方红层的地层、岩性、构造、地貌发育等进行了不同程度的研究与论述。标志着丹霞地貌从概念提出发展成为学术研究方向。

新中国成立以后，随着经济建设的需要，我国地质、地理工作者开展了广泛的区域地质调查和综合科学考察，对中国红层的分布和地层特征有了更多的了解，丹霞地貌作为一种地貌类型在各类考察报告中得到更广泛的使用。1960 年曾昭璇发表了《岩石地形学》，首次将红层地貌作为一类独立的岩石地貌类型在专著中总结论述。1961 年黄进把丹霞地貌作为广东省的一种地貌类型进行地貌填图，并首次对丹霞地貌的定义作了阐述。

“文革”期间，丹霞地貌的学术研究基本处于停滞状态。“文革”后，曾昭璇、黄少敏(1980)在《中国自然地理·地貌》一书中，专题论述了中国红层的分布、岩石学特征、地貌发育过程和形态特点，对以丹霞山为代表的主要丹霞地貌发育和地貌特征作了系统总结。丹霞地貌这一学术名词得到了更为广泛的传播与接受。

黄进从 80 年代以来主要从事丹霞地貌的研究。尤其是 1989 年以来，黄进先后得到中山大学和国家自然科学基金的资助。他主持的课题组先后对中国境内 28 个省、市、自治区的 400 多处丹霞地貌进行了考察，同时也带动了一批学者参与当地的丹霞地貌研究。

随着我国旅游发展对资源深层次开发的需要，丹霞地貌风景资源开发对丹霞地貌的基础研究和旅游开发实践研究提出了新的要求，并产生有力推动，出现前所未有的发展局面。

1991 年，由陈传康、黄进发起在丹霞地貌的命名地——广东省丹霞山召开了第一届全国丹霞地貌旅游开发学术讨论会，并成立了“丹霞地貌旅游开发研究会”。从此，学术研究异常活跃，社会影响日益扩大。在黄进等人的组织下，研究会又先后在福建武夷山、湖南崀山、青海坎布拉、湖南万佛山和贵州习水夜郎山召开了五届学术研讨会。并出版了六本论文集，参加交流的论文 230 多篇，收入论文集的 163 篇。研究内容涉及基础理论、研究方法、历史文化、开发利用和科普教育等方面，成为当代地貌学领域中一个有力的生长点。作为地貌学的一个分支，它已日渐走向成熟，并直接或间接地服务于经济建设，得到学术界与社会的关注。

During an investigation in the northern part of Guangdong, the Chinese geologist Feng Jinglan discovered that there was a significant geomorphologic contrast between the red sandy conglomerate in Mt. Danxia and the surrounding pelitic red beds, and therefore named the red beds constituting Mt. Danxia as Danxia Bed. In 1938, Chen Guoda studied the geomorphologic development features of the red beds in Mt. Daxia and established for the first time the concept of Mt. Danxia Landform. Since 1938 when Chen formally used the term Danxia Landform, other researchers have continued to use the concepts of Danxia Bed and Danxia Landform. Subsequently, many researchers have contributed to various degrees to the study of the stratigraphy, lithology, structure and geomorphology of the red beds in South China. This suggests that beginning with the initiation of the concept, Danxia Landform has developed into a research orientation.

After the founding of the People's Republic of China, to satisfy the need of economic construction, Chinese geologists and geographers have carried out extensive regional geological survey and comprehensive scientific investigation, resulting in further understanding of the distribution and stratigraphic features of China's red beds, while at the same time the Danxia Landform as one of the geomorphologic types has found a wide application in the various investigation reports. In 1960, Zeng Zhaoxuan published his *Rock Geomorphology*, in which he described and summarized for the first time red bed landform as one of the lithologic-geomorphologic types. In 1961 when carrying out a geomorphologic mapping program in Guangdong, Huang Jin used Danxia Landform as a mapping unit, and gave a definition for the first time for Danxia Landform.

During the "Cultural Revolution," academic research on Danxia Landform came to a standstill. After the "Cultural Revolution," Zeng Zhaoxuan and Huang Shaomin (1980) specially dealt in a book "Physical Geography-Geomorphology of China" with the distribution, petrographic features, geomorphologic development and geometrical features of red beds in China, and systematically summarized the major development and features of Danxia Landform represented by Mt. Danxia. The academic term Danxia Landform was thus more widely spread and accepted.

Since the 1980s, Huang Jin has devoted himself mainly to the study of Danxia Landform. Especially since 1989, he has successively gained financial supports from Zhongshan University and the State Natural Science Foundation. His research group has carried out investigations of Danxia Landform in more than 400 places in 28 provinces, cities and autonomous regions in China, and has at the same time brought along many local researchers to take part in the investigation.

The development of tourism in China has vigorously propelled the basic study of Danxia Landform, bringing it to a new stage of development.

In 1991, initiated by Chen Chuankang and Huang Jin, the First All-China Symposium on Danxia Landform and Tourism Development was convened in Mt. Danxia of Guangdong, the place after which Danxia Landform has been named, and The Danxia Landform and Tourism Development Research Society of China was established during the symposium. From then on, academic research has been very active, and its social influence has been getting wider and wider. Organized by Huang Jin and others, the Research Society has convened five symposia successively in Wuyishan of Fujian, Langshan of Hu'nan, Kanbula of Qinghai, Wanfoshan of Hu'nan and Yelangshan, Xishui of Guizhou. Involving more than 230 papers, of which 163 has been published in six volumes of proceedings. The subjects of these papers include basic theory, research method, history and culture, development and utilization, popular science education and other aspects. Danxia Landform has now become a growing point in the geomorphologic field of to day. As a branch of geomorphology, it is now approaching maturity during its course in serving directly and indirectly for the economic construction, and has thus attracted the attention of the academic circles as well as the society.

3 丹霞地貌：红色陆相碎屑岩 + 赤壁丹崖

Danxia Geomorphology: Non-marine red clastic rock plus red walls and red cliffs

1939 年，陈国达提出“丹霞地貌”名词以来的很长时期内，没有人讨论过它的定义。直到 1961 年，黄进在编制广东省地貌图时，首次对丹霞地貌的概念进行了解释。1983 年，《地理学词典》和《地质辞典》几乎同时对丹霞地貌作了大致相同的解释。但这些解释大多是对于其岩石、构造、形态的描述。到 1992 年，黄进等把定义简化为“由红色砂砾岩形成的丹崖赤壁及其有关地貌称为丹霞地貌”。

1993 年，彭华认为有必要对红色碎屑岩加上“陆相”的限制，则丹霞地貌的定义可以表述为：以赤壁丹崖为特征的红色陆相碎屑岩地貌。此外，罗成德、刘尚仁(1994)还提出对丹霞地貌的“陡崖坡”应有高度和坡度的度量指标。

For a long time since Chen Guoda initiated the term Danxia Landform, no one discussed its definition until 1961, when Huang Jin explained this concept for the first time when compiling the geomorphologic map of Guangdong. In 1983, a roughly similar explanation appeared in “Dictionary of Geography” and “Dictionary of Geology” almost at the same time. However, these explanations are limited mostly to its petrological, structural and geometric descriptions. In 1992, Huang Jin and others defined the term Danxia Landform in an abbreviated form as "the red cliffs, red walls and other relevant landforms made up of red sandy conglomerates."

In 1993, Peng Hua suggested that it was necessary to add a restrictive attribute "non-marine" to the "red clastic rocks." Thus, the definition can be expressed as "landform made up of non-marine red clastic rock and characterized by red walls and red cliffs." In addition, Luo Chengde and Liu Shangren (1994) provided lowest limits of the height and the gradient for "steep cliff slope" of the Danxia Landform.

4 顶平，身陡，麓缓：丹霞地貌最基本的形态特征

Flat top, steep face and gentle piedmont: The most typical geometrical features of Danxia Landform

黄进把近水平构造的丹霞地貌坡面，自上而下分为三种类型：

(1) **受近水平岩层面控制的层面顶坡**：这些顶坡一般呈微微上凸平缓坡面，坡度与岩层面的控制有关，同时也与山顶风化物质在湿润条件下的内摩擦角有关。

(2) **受垂直节理控制的陡崖坡**：陡崖坡主要是因为较大面积的岩块沿垂直节理发生崩塌作用形成的陡崖。这种陡崖坡是三种坡面中最重要的一种坡面。在陡崖处于相对稳定状态时，由于风化、片流侵蚀，陡崖面上会发生突出部分圆化、软岩层凹进、坡面形成平行沟槽等微地貌。

(3) **受崩积岩块内摩擦角控制的崩积缓坡**：崖麓崩积缓坡，一般较严格地受崩积岩块内摩擦角所控制，多在 30° 左右。当崩积

物较粗大时，其坡度较陡；反之则较缓。若崩积物及风化层被蚀去，则露出其下的基岩缓坡。

“顶平、身陡、麓缓”三种坡面，是丹霞地貌中最基本、最简单的坡面类型，也是丹霞地貌的基本形态特征（图1）。

缓倾斜丹霞地貌坡面，除山顶为向一方作缓倾斜的微凸形顶坡外，其他坡面特征与水平岩层构造的坡面类型基本相同。即缓倾斜顶坡四周常发育陡崖坡，而崖麓则形成崩积缓坡，形成“顶斜、身陡、麓缓”的坡面特征。

图 1　典型的丹霞地貌方山

Fig. 1　The slope surfaces of Danxia landform which formed along horizontal rock beds

Huang Jin classified the slopes of Danxia Landform related with a sub-horizontal stratification into three types, which are given in descending order as follows:

(1) The top slope controlled by sub-horizontal bedding plane: The gradients of these top slopes are controlled both by bedding plane and by angle of internal friction of the residual materials at hilltop under humid conditions.

(2) The steep-faced slope controlled by vertical joints:

This escarpment is formed by collapse of rock blocks along a large area of vertical joint plane. This is the most important of the three types of slopes. Under a relatively stable condition, due to weathering, sheet erosion etc., the protruding parts of the escarpment will become rounded and the soft rock concaved, while other microgeomorphologic features such as parallel grooves will be formed on the slope surface.

(3) The gentle colluvial slope controlled by the angle of internal friction of the colluvial rock blocks: Strictly controlled by the angle of internal friction of the colluvial rock blocks, the angle of this slope is mostly around 30°. Larger colluvial blocks give rise to a less gentle slope and vice versa. When the colluvial blocks and the residual soils are eroded off, the bedrock underlying the gentle slope will be exposed.

The above-mentioned three types of slopes characterized respectively by "flat top," "steep face" and "gentle piedmont" are the most basic and simple types of slope featuring the Danxia Landform (Fig. 1).

5 红层：丹霞地貌发育的物质基础

Red beds: The material foundation for the development of Danxia Landform

5.1 红层：一种红色陆相碎屑岩系

曾昭璇等 (1980) 认为“红层是从中生代，特别是从侏罗纪到早第三纪的陆相红色岩系”，是丹霞地貌发育的物质基础。至于它形成于什么时代，没有必要在定义中加以限制，因为在中生代以前和晚第三纪也有红层堆积；由于沉积环境的差异和后期地质作用的改造，红层的颜色可能变化于棕黄、褐黄、紫红、褐红、灰紫等偏红色调；红层是一种岩性复杂的陆相沉积，这与相对均质、致密的海相沉积有着很大的差别；说红层是碎屑岩系主要用以区别陆相化学沉积和生物沉积。

5.2 红层是封闭的内陆盆地堆积

红层的岩性、结构和构造、沉积相和古生物化石（包括植物孢粉）等各种地层研究成果显示，红层是一种典型的陆相沉积。红层中所含的可溶性盐较多，丹霞山、崀山砂岩样中 Al_2O_3、CaO、Fe_2O_3、K_2O、Na_2O 的含量均超过砂岩的克拉克值(彭华，1992；吴起俊，1994；肖自心等，1998)；许多红层中都夹有膏盐层，粤北南雄群红色泥岩、皖南休宁盆地红层、川南和甘肃等地的红层中都夹有多层的含石膏层(彭华，1992；黄可光，1998；罗成德，1998)。故而认为大部分红层堆积在封闭的、相对干燥的内流盆地环境中。而在盆地堆积的后期，随着盆地抬升、湖水外泻而淡化，但仍处于相对封闭的内陆环境。

5.3 红层的物质组成差异巨大

根据目前的研究，红层碎屑的颗粒组成差异很大，有洪积泥砾、短促河床砾石层、河床相砂砾、较长途搬运的具有沙波结构的河床砂、河流入湖三角洲前缘相的砂质层和以泥质为主的湖盆相粉砂质或淤泥质沉积等。

红层盆地边缘往往堆积巨厚的洪积相混杂泥砾，往中心渐变为洪冲积砾岩、砂砾岩与河湖积砂岩、粉砂岩或泥质岩。但在垂直剖面上，盆地外围的洪积扇前缘粗碎屑堆积区，岩石粒级的变化很大，常常表现为互层状。即砾岩、砂砾岩交互成层，夹砂岩；或以砂砾岩、砂岩为主，夹砾岩或夹薄层粉砂岩、泥质岩。而接近湖盆中心的细碎屑堆积区，岩性的垂直变化一般较小。

碎屑的砾石组成一般与其外围山地的物源一致，岩屑、砾屑是其外围物源地岩石碎屑的混合。其砂质主要是石英，含部分长石；胶结物以泥、砂为主，化学胶结物主要为硅质、钙质和铁质。湖南崀山的碳酸钙含量随外围接触地层的不同而有很大差别，在接触碳酸岩源地的地方，碳酸钙含量可达25%以上(肖自心等，1998)。

5.4 红层不一定全是红色

红层是一种偏红色的陆相碎屑堆积，这是之所以称为红层的主要标志。但是，在一套红色岩系中，不论是垂直方向还是水平方向，都不可能保持沉积环境在时间和空间上的完全一致。加上在成岩过程中的物理化学变化和隆起后各种环境因素等后期影响，其颜色也可能再发生一些变化。因此，各地的红层在色度上有较大的差别；同一套红色岩系中上下的颜色也可能存在较大的差别，有时可能含有多层的非红色夹层。只要在一个沉积单元的一套陆相碎屑岩系的总体色调呈现偏红色，我们则没有必要因为其含有一些非红色夹层而把它排斥在红层之外。

此外，红层砾石的原岩成分、砾石本身的颜色、砾石的含量、胶结物成分、砂质、岩层的含水量等都对红层的颜色和色调产生一定的影响。

5.5 红层的岩性、岩层结构和构造变化剧烈

红层基本上由陆相碎屑组成，包括角砾岩、砾岩、砂砾岩、砂岩和细粒碎屑岩 (粉砂岩和粘土岩) 。此外，时有化学成因和生物成因的岩层 (如透镜状薄层石灰岩、泥灰岩、白云岩、石膏等) 夹杂其中。如甘肃的临夏组 (N_2l) 综合剖面中，砖红色泥岩、粘土岩互层，夹石膏及薄层灰岩 (黄可光，1998) 。此外，红层还可能包含火山堆积 (如广东三水盆地) 或风成沉积 (如四川盆地南部) 。

相对说来，洪水与河流淡水沉积、泥砂胶结或硅质、铁质胶结的砾岩和砂砾岩比较坚硬；而盆地中心沉积的粉砂岩、泥质岩所含的可溶性胶结物质较多，透水性较差、含水较多，因而比较软弱。例如，丹霞山砂岩、砾岩在水饱和时的抗压强度为441~546 kg/cm^2，而泥质砂岩仅为300 kg/cm^2 (黄进，1993) 。根据目前的研究，大部

分丹霞地貌发育在相对坚硬的砾岩、砂砾岩、砂岩的地层组合上，而相对软弱的粉砂质和泥质岩多发育红层丘陵，只有在河流凹岸或质地稍坚硬时才形成尺度不大的红色陡崖坡。

大部分粗碎屑红层以厚层或巨厚层甚至块状构造为主；接近盆地中心的湖积相或河漫滩相细碎屑红层一般以薄层构造为主；而介于两者之间的河床相和三角洲相堆积则较为复杂，比较稳定的较大河流在河床侧向摆动过程中，可能会形成厚层或巨厚层砂质堆积，同时在边滩发育或河底沙波移动过程中，在沉积层内形成斜层理或交错层理；另一种情况可能是在入湖三角洲推进过程中，在同一砂岩层内，发育斜层理或交错层理。丹霞山锦石岩段和白寨顶段有8~20 m 巨厚层不等粒石英砂岩和长石石英砂岩，均具有大型板状交错层理。据罗成德研究，四川盆地南部白垩统的巨型斜层理十分发育。层系厚 3～5 m，有时可达数十米。 细层层理倾角一般大于30°，个别 40° 以上，呈多向倾斜。

5.6 红层形成的古地理环境

地质历史上，红层沉积一般发生在古热带和亚热带盆地中，相当于古纬度 15° ~30° 范围内。由于红层中大多含有碳酸盐或钙质胶结物，甚至含有蒸发盐夹层，大部分学者认为红层形成在炎热干燥的环境下。但是红层的红色主要是高价铁(Fe^{3+})相对富集而成的，这个富集过程必须要有足够的淋溶作用。因此，红层形成的古气候环境应是一种类似于现代热带、亚热带半湿润、半干旱气候或干湿季气候，以保证有足够的淋溶并保持长时期的氧化环境。

(1) Red beds—a nonmarine red clastic rock series: Zeng Zhaoxuan *et al.* (1980) holds that "red beds are non-marine red rock series formed in Mesozoic times, specifically from the Jurassic to the Early Tertiary." They are the material foundation for the development of Danxia Landform. The geologic age in which they were formed is not important for the definition because deposition of red bed also occurred in pre-Mesozoic times and in the Late Tertiary. Owing to difference in depositional environments and influences of late stage geologic processes, the color of red beds can change into brownish reddish yellow, brownish yellow, purplish red, brownish red, grayish purple and other reddish tints. Red bed is a nonmarine deposit complex in lithologies, and is significantly different from marine deposit which is relative homogeneous and compact. It is a clastic rock series, which is distinguishable from other nonmarine deposits such as the chemical deposits and biogenic deposits.

(2) Red beds—a deposit in closed continental basin: Based on stratigraphic studies on lithologies, texture, structure, sedimentary facies, and fossils (including spore-pollen), red beds are a typical continental deposit, which contains considerable soluble salt. Analysis of sandstone samples from Mt. Danxia and Liangshan has revealed that the contents of Al_2O_3, CaO, Fe_2O_3, K_2O, and Na_2O exceed the corresponding clarke values for the average sandstone (Peng Hua, 1992; Wu Qijun, 1993; Xiao Zixin *et al.*, 1998). Many red beds have gypsum-salt intercalations: for example, multilayered gypsum is found intercalated in red mudstone of the Nanxiong Group in North Guangdong, in red beds of the Xiuning basin in South Anhui, and in red beds in South Sichuan and Gansu (Peng Hua, 1992; Huang Keguang, 1998; Luo Chengde, 1998). Therefore it is believed that most of the red beds were accumulated in closed, relatively dried continental basin environment. During the late stage of basin accumulation, however, with the uplifting of the basin and the out flowing of lake water, desalination will have occurred although the basin still remained in a relative closed inland environment.

(3) Varied constituents of red bed: Based on recent studies, red beds are very contrast in clastic composition. They may consist of pluvial mud gravel, intermittent river bed gravel, fluvial sandy gravel, river bed sands which display sand ripple and which experienced long-distance transportation, sandy layer in lake-inlet delta front, and pelitic-dominant lake basin silty or mucky deposits.

On the margin of the red basin are often accumulated pluvial chaotic mud gravels of tremendous thickness, which gradate toward the center into pluvial-alluvial conglomerate and sandy conglomerate, and fluvial-lacustrine sandstone, siltstone or pelitic rocks. In front of a torrential fan at the periphery of a basin where coarse clastic accumulations are formed, however, grain sizes of the rock are highly variable vertically. This is often demonstrated by interbedding of conglomerate and sandy conglomerate intercalated with sandstone, or by sandy conglomerate and sandstone intercalated with conglomerate or thin beds of siltstone and pelitic rock. Nevertheless, approaching the center of the lake basin where fine clastic rocks are accumulated, vertical variation in lithologies is often insignificant.

The composition of the gravels is as a rule in conformity with the provenance in a mountainous area around the basin. The lithic and rudaceous fragments are the mixture of debris from the above-mentioned provenance. The sand grains are dominated by quartz, with partly feldspar, while the cementing materials consist largely of mud and sand, and the chemical cementing materials are mainly siliceous, calcareous and ferruginous substances. The calcium carbonate contents in Liangshan, Hunan vary greatly with the contacted strata at the periphery: when the contacted strata consist of carbonate rocks, the calcium content can be as much as more than 25% (Xiao Zixin *et al.*, 1998).

(4) Red beds are not necessary red: Red beds are a nearly red colored continental clastic accumulation, and red color is the main criterion by which the red bed is named. Nevertheless in a suite of red rock series, depositional environment cannot remain the same either vertically or horizontally. Moreover, physical-chemical change in diagenetic stage and various environmental factors after uplifting will cause their color to change further. Therefore the tone of the red bed is different from place to place. In one and the same suite of red rock series, the color can be different greatly from its upper part to its lower part. Sometimes, this suite can contain many layers of non-red intercalations. However, if the overall tone of a suite of continental clastic rock series appears to be nearly red, we need not exclude it from the red bed family even though it contains some non-red intercalations.

In addition, the composition, color and content of gravel in the red bed, the composition of the cementing material, the sandy substances, and the water content of the stratum will all affect the color and tone of the red beds.

(5) The lithology, rock texture and structure of red bed are highly variable: Red beds consist basically of continental clastic rocks, including breccias, conglomerates, sandy conglomerates, sandstones and fine-grained clastic rocks (siltstones and clayrocks). In addition, there are occasionally some intercalations of chemically formed rocks and biogenic rocks such as marls, dolomites and gypsums. For example, in a comprehensive cross section of the Linxia Formation (N_2l) in Gansu, the interbedding brick red mudstone and clayrock are intercalated with gypsum and thinly bedded limestone (Huang Keguang, 1998). Furthermore, red beds can also contain volcanic accumulations (e.g. in the Sanshui basin of Guangdong) or aeolian deposits (e.g. in south part of the Sichuan basin).

Generally speaking, pluvial and fluvial deposits, and conglomerate or sandy conglomerate cemented by muddy sand, siliceous or ferruginous materials are relatively hard, but siltstones and pelitic rocks deposited in the central part of the basin are relatively weak because they contain abundant soluble cementing materials and are poor in permeability but rich in water content. For example, when saturated with water, the sandstone and conglomerate from Mt. Danxia have a compressive strength of 441~546 kg/cm^2, while that of pelitic sandstone is as low as 300 kg/cm^2 only (Huang Jin, 1993). According to recent studies, Danxia Landform is developed mostly in relatively hard strata assemblage made up of conglomerate, sandy conglomerate and sandstone, while the ordinary red bed hills are developed mainly in relatively weak strata composed of silty or pelitic rocks. Only on the concave bank of a river or when the rocks are hard can a small-scaled red escarpment slope be formed.

Most of the coarse clastic red beds are dominated by thickly bedded, very thickly bedded and even massive structures; lacustrine or flood land fine clastic red beds approaching the center of the basin are usually dominated by thinly bedded structure; while those distributed between the above-mentioned two kinds of red beds are the riverbed and deltaic deposits, which are relatively complex. As to the riverbed deposit, when the river is large and stable, thickly or very thickly bedded sandy deposit can be formed during the laterally swinging of the riverbed; at the same time, oblique bedding or cross bedding can be formed in the sedimentary stratum during the development of concave flood plain or during the migration of alluvial sand wave. On the other hand, with regard to the deltaic deposit, oblique bedding or cross bedding can also be formed in sandstone when the lake inlet delta pushes its way forward. In Mt. Danxia, large-scale tabular cross beddings are recognizable both in very thickly bedded inequigranular quartzose sandstone and feldspar-quartz sandstone, measuring 8~20 m thick within the Jinshiyan Member and the Baizhaiding Member. According to Luo Chengde, large-scale cross beddings are also well-developed in the Cretaceous Series in the southern part of the Sichuan Basin: the stratum is 3~5 m thick, with its cross bedding dipping in many directions generally at greater than 30° and individually at >40°.

(6) The palaeogeographic environment in which red beds were formed: In the geological history, deposition of red bed took place generally in palaeotopical or palaeosubtopical basins, extending between palaeolatitudes 15° and 30°. Since most of the red beds contain carbonate or calcareous cementing materials and even evaporate intercalations, most of the researchers believe that red beds were formed in a hot and arid environment. However, the red color of red bed is responsible largely by the enrichment of ferric ions (Fe^{3+}), and this enrichment must be aided by sufficiently leaching process. Therefore the palaeogeographic environment in which red beds were formed should be similar to a tropical, subtropical semi-humid and semi-arid climate or an alternating dry and wet seasonal climate, in order to maintain a sufficiently leaching process and a long-term oxidation condition.

6 地壳演化的特定历史产物：足够大的陆地出现之后——红层的形成时代

Secific historical products of crustal evolution: The age in which red beds were formed came after the appearance of land sufficiently large in dimensions

根据国内学者的研究，中国目前所发现的发育丹霞地貌的红层均不早于中生代，其中以白垩系红层最多，约占 80% 。形成丹霞地貌的最老红层为三叠系 (陈致均，1994；罗成德，1996)，还没有发现丹霞地貌发育在更老的地层中。

但这并不说明中生代以前没有红层堆积。其实，北美—俄罗斯古陆上 (主要在北欧、格陵兰岛到北美洲东北部) 加里东运动期间就已经在山前拗陷和山间盆地中堆积了陆相“老红层”(老红砂岩，Old Red Sandstone)。阿巴拉契亚地槽太康运动 (O_3) 第二幕之后，褶皱隆起带(地背斜)以西与劳伦古陆之间的拗陷中，沉积了很厚的陆相红色页岩、砂岩和砾岩等碎屑岩。到了晚志留世，红层堆积向内陆扩展，美国中北部的几个盆地中均广泛发育了红层和蒸发岩 (Alling & Briggs, 1961)。晚古生代，在北美—俄罗斯古陆的大部分地区和冈瓦纳古陆的北部 (包括澳洲)，都曾有红层和蒸发岩发育。中国塔里木地台在泥盆纪和二叠纪也发育红色陆相粗碎屑堆积 (王鸿祯、刘本培，1980)。因此，红层开始发育的时代应前推至早古生代（图2)。

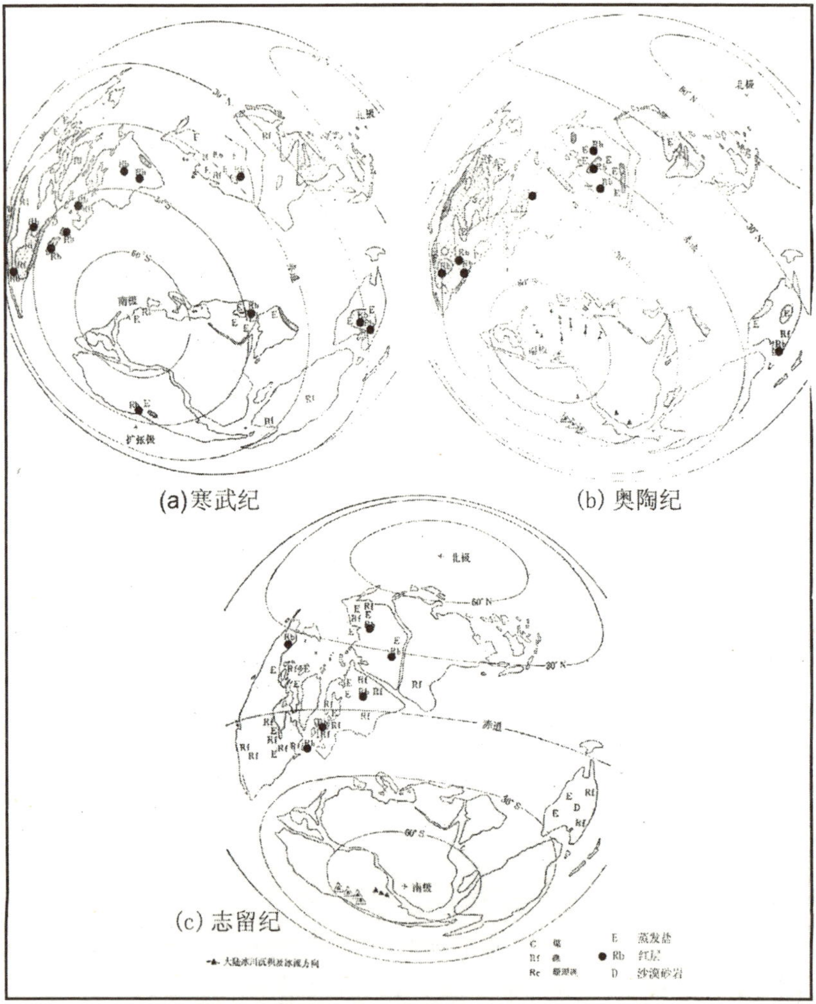

图 2　古生代世界红层分布（据 Seyfert & Sirkin，引自白顺良等）
Fig. 2　World distribution of Paleozoic red beds
(after Seyfert & Sirkin and Bai Shunliang *et al.*)

彭华 (1994) 认为红层是地壳演化到一定历史阶段的产物，即地壳刚性增强，地台扩大，有相对广大的陆地面积之后，才因地台活化或褶皱山地围闭而形成反差较大的内陆沉积盆地和外围山地物源区，在适当的气候与沉积环境下发育陆相红层堆积。

地壳的演化史是一个有硅铝质的大陆壳逐渐分异并聚集加厚的过程（彭华，1985）。在这个过程中，陆壳通过反复地地质旋回而固化，直到元古代后期形成了相对完整的几个古陆台。也只有在这种背景下，才有可能在一些古陆台上出现相对封闭的内陆盆地，在气候适宜的情况下形成红色陆相沉积。因此，目前所发现的红层和蒸发岩系出现于早古生代古陆台的山间盆地或地台拗陷中。

According to Chinese researchers, red beds constituting Danxia Landform found so far in China are all not older than the Mesozoic, and about 80% of them were deposited in Cretaceous times. The oldest red beds that constitute Danxia Landform are deposited in the Triassic Period (Chen Zhijun, 1994; Luo Chengde, 1997). Danxia Landform developed in red beds older than the Triassic has not been found yet.

However, this does not mean that no red beds were formed prior to the Mesozoic Era. In fact, in North America-Russian Old Land (specifically from North Europe, Greenland Island to the northeast part of North America), continental Old Red Sandstone was accumulated in foredeep and intermontane basin during the Caledonian Movement. In Appalachian geosyncline, a very thick suite of continental red clastic rocks including shale, sandstone and conglomerate were deposited in the depression between the west of a upwarping folded belt (geoanticline) and the Laurentian Old Land subsequent to phase II of the Taconian Orogeny (O_3). In the Late Silurian, accumulation of red beds extended toward the inland. As a result, red beds and evaporates are widely distributed in several basins in the central-north part of the United States (Alling and

Briggs, 1961). In Late Palaeozoic times, red beds and evaporates were formed in most of the North America-Russian Old Land as well as in the north part of Gondwana Old Land (including Australia). In Tarim Platform in China, continental red coarse clastic rocks are also found in the Devonian and Permian Systems (Wang Hongzhen and Liu Benpei, 1980). Therefore, the age when red beds began to develop should be as early as the Early Palaeozoic Era (Fig. 2).

Peng Hua (1994) believes that red beds were formed only when the Earth's crust had evolved into a specific historical stage, in which the crust had increased in rigidity, the platform had been enlarged, and the land vast enough to produce the highly contrast inland sedimentary basins and the peripheral mountainous provenance due to rejuvenation of platform or due to enclosure by folded mountains. Only in such environment and when the climate was appropriate could the continental red beds be accumulated in the basins.

The evolutional history of the Earth's crust is a process when the continental crust containing sialic materials gradually underwent differentiation, accumulation and became thickening (Peng Hua, 1985). During this process, the continental crust consolidated repeatedly through geologic cycles until the Late Proterozoic Era when several relatively complete palaeoplatforms were formed. Only in this background can the relatively closed inland basins be formed in some palaeoplatforms, and begin to receive continental red sediments under favorable climatic conditions. Hence, red beds and evaporate series found today occur in Early Palaeozoic intermontane basins of palaeoplatform or in platform depressions.

7 地质构造：控制着丹霞地貌的发育

Geologic structure: Controlling the development of Danxia Landform

7.1 区域构造对沉积盆地的控制

中生代以来，我国许多在海西运动以后稳定的地台发生强烈活化。东部地区受太平洋板块活动的影响，形成一系列北东向和北北东向的隆起带与拗陷带（图3）；西部地区则受印度板块影响，从北向南渐次成陆并形成若干盆地；中部则形成一个北东向的压扭性地带。因此,中国中、新生代盆地基本以此格局展布，控制了红层盆地的分布规律。

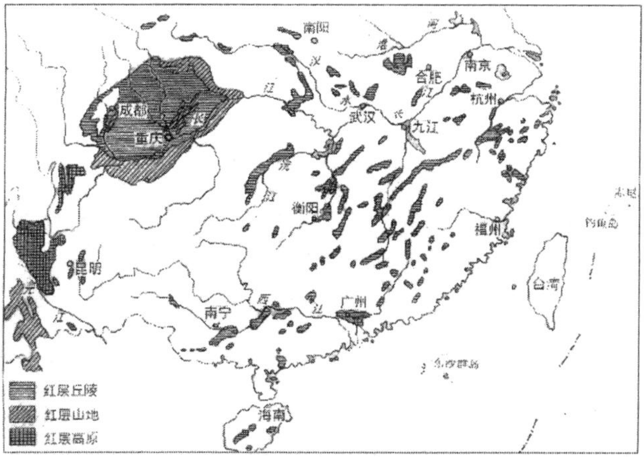

图 3　中国东南部红层分布 (据曾昭璇等，1980)

Fig. 3　A brief distribution of red beds in southern China (After Zeng *et al.*, 1980)

7.2 断层节理对山块格局的控制

盆地内部的构造线格局是控制丹霞地貌山块格局乃至山块形态的基本因素。大的构造线控制了山块总体的排列方向，小构造则控制山块的走向、密度和平面形态。丹霞山的山块排列基本沿北北东向的大断层延伸，而山块的走向、石柱的排列主要沿近东西向的断层和大节理延伸 (图4)。

图 4　断层节理对丹霞地貌山块的控制（丹霞山天柱石一带）

Fig. 4　The relationship between tectonic lines and mountain blocks arrange displayed by aerial photography (near by Tianzhushi area, Mt. Danxiashan)

7.3 岩层产状对坡面形态的控制

根据黄进等人 (1982, 1992) 的研究，岩层产状对丹霞地貌形态的影响主要是对于山块顶面和构造坡面的控制。一般情况下，近水平岩层上发育的丹霞地貌具有“顶平、身陡、麓缓”的坡面特征；缓倾斜岩层上发育的丹霞地貌则“顶斜”，具有单面山的特点（图5），其斜顶基本和岩层层面一致；陡倾斜岩层所发育的丹霞地貌若不是保留了古侵蚀面的话，其顶面很难形成平顶或缓斜顶，而往往是尖顶。甘肃刘家峡地区的岩层倾角达50° ~ 60° (黄可光，1992)，其构造坡面已构成陡崖坡。

7.4 地壳升降对地貌发育进程的控制

地壳升降的影响体现在红层盆地必须是后期上升区，以便为侵蚀提供条件。上升到一定程度而长期相对稳定，有利于丹霞地貌按连续过程从幼年期到老年期逐步演化；间歇性抬升则可能发育多层性丹霞地貌，安徽齐云山的这种陡缓坡组合多达五级（图6）。据黄进、刘尚仁等 (1994) 在丹霞山区河流阶地冲积层中进行的热释光采样分析，得知丹霞山区的地壳平均上升速度为 0.97 m/万年。由此可知，丹霞山现代地貌形成于距今大约600万年前， 其丹崖后退速度平均为0.5～0.7 m/万年。

(1) The controlling effect of regional structure on sedimentary basin: From the Mesozoic onward, strong activation took place in many platforms that had been stable after the Hercynian orogeny in China. In the eastern area, a series of NE- and NNE-trending upwarping zones and down-warping zones were formed under the influence of Pacific Plate activity(Fig. 3); in the western area, land emerged gradually southward and some basins were formed under the influence of the India Plate; in the central area, however, a NE-striking compresso-shear belt took shape. Therefore, the Mesozoic and Cenozoic basins in China are distributed basically in such a pattern, which controls also the distribution of the red basins.

(2) The controlling effect of faults and joints on hilly block distribution framework: The structural line framework within the basin is the basic factor controlling the distribution pattern of the hilly blocks of Danxia Landform, and even the geometry of the hilly blocks. Major structural lines control the overall direction in which the hilly blocks are arranged, while minor ones govern the strike, density and planar outline of the hilly blocks. The hilly blocks of Mt. Danxia are aligned essentially along NNE-striking major faults, while the strikes of the hilly blocks and the alignment of the hoodoo columns are in conformity with that of the E-W fault and major joints (Fig. 4).

(3) The controlling effect of the attitude of strata on the geometry of the slope: According to Huang Jin and others (1982, 1992), in Danxia Landform, the attitude of strata controls mainly the top surface and structural slope of the hilly blocks. Generally, a Danxia Landform developed in nearly horizontal strata is characterized by "flat top, steep face, and gentle piedmont;" that in gently dipping strata is characterized by "oblique top," showing the features of a cuesta(Fig.5), with its oblique top surface essentially in conformity with the bedding plane; while in case with a Danxia Landform developed in steeply dipping strata, if the palaeo-erosional surface is not preserved, pointed top is often formed because it is difficult for the top surface to evolve into a flat top or a gently dipping top. For example, in the Liujiaxia area of Gansu where the strata dip at angles as much as 50°~60° (Huang Keguang, 1992), the structural slope has already evolved into an escarpment.

图 5　发育在 25° 岩层上的丹霞单面山地貌

Fig. 5　Danxia cuesta landform developed in 25° -inclined rock bed

(4) The controlling effect of vertical movement of Earth's crust on the development process of landform: This effect is demonstrated in that the red basins must be an uplifting area in late stage in order to provide the necessary conditions for erosion to take place. After uplifting to a certain level, the crust must maintain relatively stable for a sufficiently long period to facilitate Danxia Landform to evolve gradually and continuously from young to old stages; intermittently uplifting, however, can give rise to multi-terrace Danxia Landform. In Qiyunshan of Anhui, for example, this kind of alternating steep and gentle slopes association can have as many as five terraces(Fig. 6). According to thermoluminescent dating made on samples from alluvium of the river terrace of Mt. Danxia area by Huang Jin, Liu Shangren *et al.* (1994), the Earth's crust of Mt. Danxia area has been uplifted at a rate of 0.97 m/10000a on the average. Hence, the recent landform of Mt. Danxia was formed at about 6 Ma B.P., while the red escarpment retreats at a rate of about 0.5~0.7 m/10000a on the average.

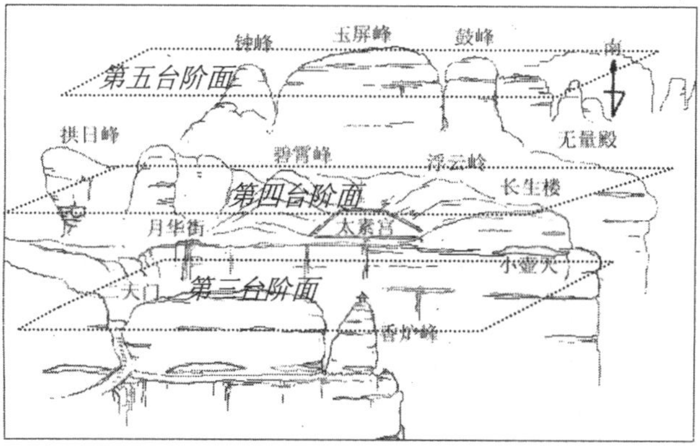

图 6　安徽齐云山阶梯状地貌(上部三层素描)

Fig. 6　Multiple layers landforms in Qiyunshan Mountain (only depict the upper two layers)

8 外力条件：丹霞地貌发育的雕塑师

External agency: The sculptor of Danxia Landform

直接影响丹霞地貌发育的外动力因素主要有流水、风化和重力等作用，其中流水是塑造地貌的主动力。在干旱区，风力对于外表形态的塑造具有不可忽视的作用；在湿润区，生物对于风化作用具有一定的影响。

8.1 流水作用

流水作用在丹霞地貌发育和演化中的主导性表现为流水是下切和侧蚀的主动力(图7)；同时流水又不断地蚀去坡面上的风化物质，使风化得以继续进行；流水的侧蚀往往在坡脚掏出水平洞穴，使上覆岩块悬空，为重力崩塌提供了可能。此外，流水对红层中的可溶性成分进行溶蚀，可促进水动力侵蚀的加强和风化作用的进程。

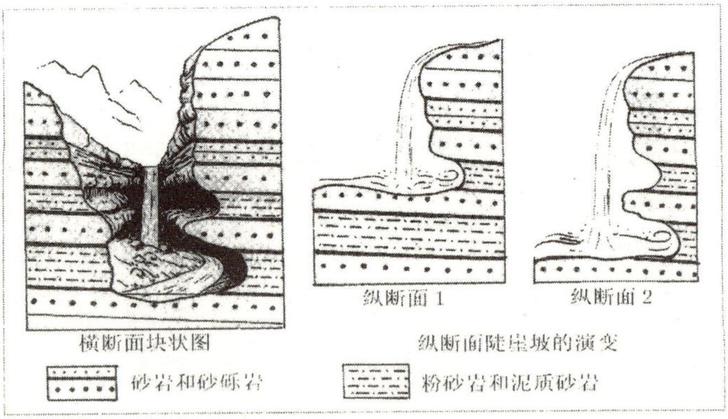

图 7　流水的下蚀与侧蚀

Fig. 7　Down-cutting and lateral erosion of flowing water

8.2 风化作用

风化作用对暴露的红层坡面进行着经常性的破坏，尤其在陡崖坡上，流水的作用减弱，在一些直立坡或反倾坡上基本无流水作用，各种风化作用 (片状、块状、粉末状等) 表现得十分清楚。因为红层在垂向上的岩性差异而导致抗风化能力的不同，常使得砂砾岩等硬岩层相对凸出而成顺层岩额或岩脊，而泥质或粉砂质软岩层则凹进而成顺层岩槽或顺层岩洞，形成丹霞地貌陡崖坡上独特的微地貌景观（图 8）。

图 8　岩层差异风化形成的顺层槽脊

Fig. 8　Ravine-and-ridge along bedding formed by differential weathering of rock

黄进(1996)认为由于红层孔隙度大、矿物成分复杂，导温性能差，尤其易于片状风化剥落。他把红层的风化作用归为凸片状风化和凹片状风化两种，前者使山顶、山脊或石块圆化，后者使软岩层凹进。并认为凹进岩槽的某一部分可继续风化发育成扁平洞，进而发育成穿洞，部分穿洞可继续风化及崩塌，发展成为天生桥（图 9，图 10）。

此外，干旱区的盐风化、高寒区的冻融风化使这些地区的丹霞地貌物理风化强烈，而使其形成比较粗糙的表面。

图 9 浙江烂柯山天生桥
Fig. 9 Natural Bridge in Mt. Lanke of Zhejiang

图 10 广东丹霞山天生桥
Fig. 10 Natural Bridge in Mt. Danxia of Guangdong

8.3 重力作用

因为陡崖坡往往是崩塌面或经后期改造过的崩塌面，是丹霞地貌最具特色的形态要素，所以重力作用在丹霞地貌发育过程中相当重要。重力作用往往发生在流水下切或侧蚀而形成的临空谷坡上，当流水侧向掏蚀而使山坡局部悬空时，悬空岩体便可能沿原生构造节理或减压（卸荷）节理发生崩塌（图 11）。此外，陡崖坡上的风化凹槽进一步加深，上覆岩体失去平衡也可沿破裂面发生崩塌（图 12）；洞穴、天生桥的顶板也常因风化而发生局部崩塌等等。

陡崖坡的崩塌大多是沿着某一破裂面的块状崩塌，到坡脚发生机械破碎，因而坡脚常堆积由巨大石块构成的崩积物（图13）。例如，丹霞山锦石岩陡崖下的崩积物最大的可达30 m×20 m×5 m，丹霞山宾馆在上面建了一幢三层楼房。

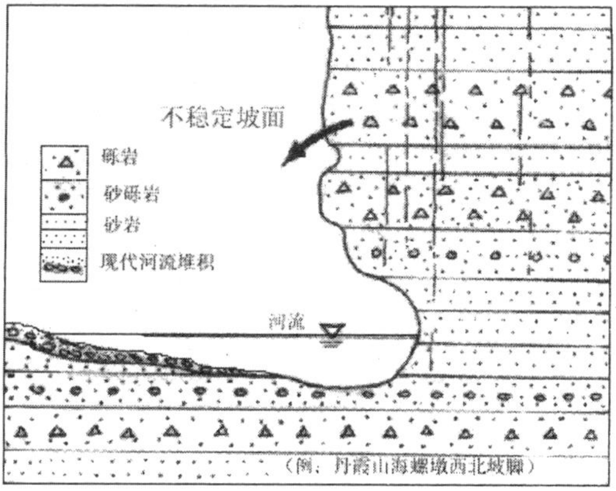

图11　流水侧向侵蚀而引起的坡面不稳定现象

Fig. 11　Unstable slope caused by lateral erosion of flowing water

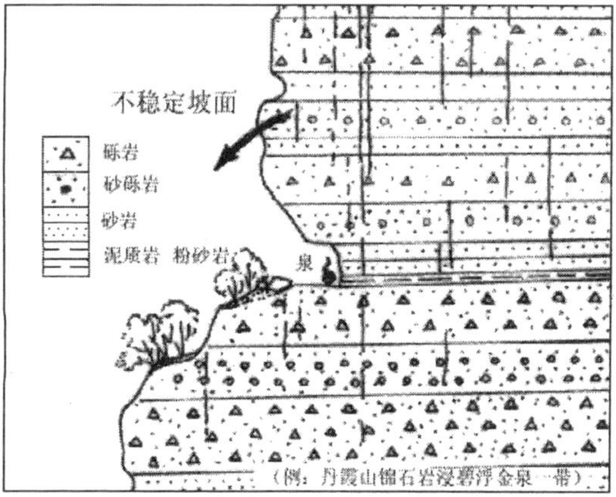

图12　主要由软岩层风化而引起的坡面不稳定现象

Fig. 12　Unstable slope caused mainly by differential weathering of the weak rock beds

8.4 其他外动力

黄进 (1991)、黄可光等 (1992)曾论述了干旱和半干旱区红层由于盐风化作用而使岩层粉化的现象，以及风沙吹磨而在丹霞崖壁上形成大量的风蚀窝穴，黄可光称之为蜂窝状丹霞地貌；黄进等 (1994) 论述了海岸丹霞地貌发育过程中波浪的作用；人工丹霞地貌是指人工凿石而在本已夷平的红层丘陵上重新雕凿出的各种地貌景观，以广东莲花山最具有代表性 (徐君亮等，1998)。

The exogenic processes responsible for the development of Danxia Landform mainly include flowing water, weathering and gravitational processes, of which the flowing water process is the main factor in sculpturing the landform. In arid regions, wind abrasion plays an significant role in sculpturing the outward appearance of landform; in moist regions, however, biological weathering become more or less active.

(1) Fluvial process: During the development and evolution of Danxia Landform, the role of fluvial process is demonstrated in its down-cutting and lateral erosion(Fig. 7); at the same time, flowing water can constantly carry away the decomposed materials on the slope surface to facilitate weathering to continue; lateral erosion of flowing water will often give rise to the formation of horizontal caverns at lower footslope, leaving the rock above to hang overhead, thus providing the possibility for gravitational collapse to occur. In addition, flowing water can dissolve soluble constituents in the red beds, so as to accelerate the hydraulic erosion and weathering process.

(2) Weathering: Weathering will constantly destruct the slope surface of the exposed red beds. Especially in steep escarpment where flowing water process weakens, and in some slopes that are vertical or inclined inward, where fluvial process is basically absent, various weathering processes (sheet-like, block-like and powder-like) are manifested very distinctly. Due to the lithologic difference (and hence the difference in resistance to weathering) of red beds in vertical direction, competent rocks such as sandy conglomerate often protrude above other weaker rocks to form "rocky forehead" or rock ridge along stratum, while incompetent rocks such as pelitic or silty rocks may become concave to form rock tanks or rock holes along bedding, resulting in a peculiar micro relief landscape on escarpment slope surface of Danxia Landform (Fig. 8).

Huang Jin (1996) believes that red beds are susceptible to sheet weathering and spalling because of their high porosity, complex mineral composition and poor temperature conductivity. He has recognized two types of weathering in red beds, i.e. convex sheet weathering and concave sheet weathering. The former will render the hilltop, ridge or rock block rounded while the latter will cause the weak strata to cave in. He also believes that a certain part of the caved-in rock tank can be weathered into pancake-shaped cavern, which will develop further into perforated rock, and finally, through further weathering and collapse, into natural bridge(Fig. 9,10).

In addition, salinization in arid region and freeze-and-thaw action in paramos region will strengthen the physical weathering of Danxia Landform in such regions, resulting in the formation of a relatively rough surface.

(3) Gravity: Gravity plays a vital role in the development of Danxia Landform because the most characteristic geometric element of Danxia Landform, i.e. the escarpment slope is often formed by avalanche. Gravity process often occurs when a valley slope caves in locally through down-cutting or lateral erosion of the flowing water. The overhanging rock blocks thus formed in the caved-in part may collapse along primary joints or stress release joints (Fig. 11). Furthermore, when the weathering tough on the escarpment slope deepens further, the overhanging rock masses may lost balance and collapse along various fracture planes (Fig. 12). The tops of caves and natural bridges etc. may also collapse due to weathering.

Collapse in escarpment slope is often a block collapse occurring along a certain fracture plane. The collapsed blocks falling on the toe of the slope will undergo mechanical disintegration. Therefore colluvial deposit consisting of huge rock blocks is often accumulated at the toe (Fig. 13). For example, under the Jinshiyan escarpment of Mt. Danxia, the largest collapsed rock block reaches 30 m×20 m×5 m in volume. The Mt. Danxia Hotel has built a three-story building on this huge rock block.

图 13 丹霞山下的崩积岩块
Fig. 13 Colluvial rock blocks on the foot of Mt. Danxia

(4) Other exogenic processes: Huang Jin (1991), Huang Keguang *et al.* (1992) have described the pulverization phenomenon of red bed in arid and semi-arid regions due to salinization, and also the large quantity of blowholes (which are referred to as alvolar Danxia Landform by Huang Keguang) on the Danxia cliff wall due to the abrasive effect of the sand-carrying wind. Huang Jin *et al.* (1994) have dealt with the wave action during the development of Danxia Landform on the coast area. Artificial Danxia Landform refers to various geomor-phologic landscapes sculptured by hand on the red bed hills, which is most typically represented by Lianhuashan in Guangdong (Xu Junliang *et al.*, 1998).

9 丹霞地貌发育的基本过程：红层盆地抬升，流水下切，边坡崩塌后退

Fundamental process in the development of Danxia Landform: Uplift of red basins, down cut by flowing water, slope collapse and retreat

黄进(1982)曾首次系统论述了丹霞地貌坡面发育基本规律，说明了它是一种不同于其他地貌的特殊地貌过程。笔者根据近年来各家的研究，将丹霞地貌的发育过程描述如下（图 14）。

丹霞地貌的发育开始于红层盆地的抬升。由于红层盆地发育在相对刚性的陆台上，因此,红层盆地的后期构造变动往往以断裂块状构造为主。红层盆地抬升的同时，其原始洼地往往继承性发育主河谷；断层破碎带和大节理成为流水切割的薄弱地带，也可能发育主河谷。主河谷两侧高出当地侵蚀基准面的红层顶面首先被流水沿断层和垂直节理下切侵蚀，形成深狭的切沟，进一步侵蚀使之加深、加大而成一线天式的巷谷。在此阶段，除主河谷两坡之外，切沟或巷谷的谷坡崩塌一般不发育。

巷谷型支谷在流水下切到一定深度，遇到下伏硬岩层或是接近于局部侵蚀面时，水流以侧向侵蚀为主对谷壁的基部进行破坏，巷谷的谷壁沿垂直节理逐步崩塌而使巷谷以加宽为主形成峡谷。这时崩塌可能在沟谷的谷底堆积，发育不稳定的崩积缓坡。随着谷地的加宽，崩积缓坡也越明显。

此后，陡崖坡基部因为流水侵蚀作用的减弱或很少有流水直接作用，陡崖坡的崩塌主要靠软岩层的风化凹进而缓慢进行。黄进(1982)认为，崩积物掩埋的崖麓基岩可暂时被保护，在陡崖坡崩塌

后退的过程中，崖麓下方造成崩积缓坡的同时，也在崩积物下面形成一个基岩缓坡面；上述作用继续进行，陡崖坡则不断崩塌后退，崩积坡不断加宽加高，其下伏的基岩缓坡面也不断加宽加高。

随着谷坡的崩塌后退，山顶的平缓坡面则被切割，并使其面积逐渐缩小；同时使山麓缓坡逐渐扩大，形成面积较大的缓坡丘陵。原来的山块则退缩成为“堡状残峰”或成为孤立的石柱。上述过程可能随地壳的间歇性抬升或流水的间歇性下切而在一个地区多次重复进行，所以在丹霞地貌区常可见到数级陡缓坡阶梯式地貌（因岩层的软硬差异，也可形成多级陡缓坡面）。

此外，根据黄进的研究，当陡崖坡发生崩塌时，巨大的重力使崩塌下来的岩块发生一次剧烈的碎化作用。这样就使堆积在崖麓缓坡上的岩块、岩屑大大增加了与空气、水分和生物的接触面积，使风化作用得以加速进行。而且，当陡崖坡发生一次规模较大的崩塌作用之后，常间以一个稳定时期。这不但使陡崖上的片状、块状风化剥落作用得以进行，而且使崖麓缓坡的崩积物有一个较安定的时期来进行风化作用，使崖麓缓坡上有相当一部分形成红色风化壳及土壤层。这时坡面水流也在缓坡上进行冲刷、侵蚀。当缓坡上某些部分的土层被蚀去，而露出基岩缓坡面时，水流就会沿着岩层的节理进行侵蚀而形成新的陡崖坡(也可以发生在地壳重新抬升的背景下)。然后又在其麓部产生新的崩积缓坡及其下伏的基岩缓坡面，并随着其上方陡崖的崩塌后退而不断扩大。

Huang Jin (1982) systematically dealt for the first time with the fundamental law governing the slope development of Danxia Landform, suggesting that it is a peculiar geomorphologic process different from other landforms. Based on the recent studies of the various researchers, the writer summarizes the development process of Danxia Landform as follows (Fig. 14).

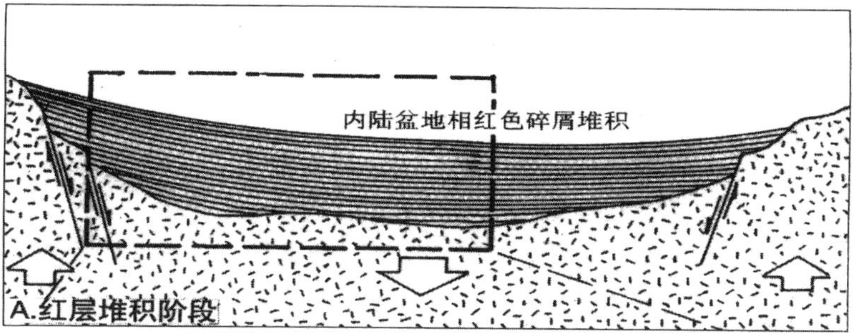

14-A. 内陆盆地形成，氧化环境，碎屑堆积（洪积、冲积、湖积）
The formation of inland basin; oxidation environment; accumulation of detrital matters (deluvial, alluvial and lacustrine accumulation)

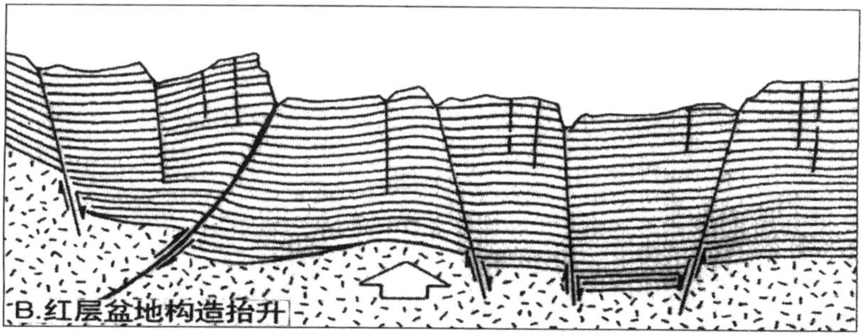

14-B. 盆地抬升，以断裂为主的块状构造发育，局部宽缓褶曲，地壳渐趋稳定
Uplift of basin and development of faulted blocks and local wide and gentle fold, the Earth's crust approaching gradually stability

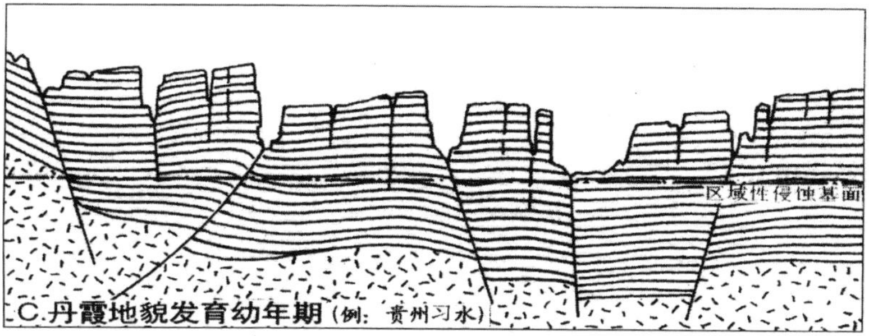

14-C. 流水下切，巷谷、峡谷发育，上部保持较大面积的沉积顶面或弱侵蚀平台
Flowing water mainly down-cutting, development of tunnel-like valley and canyon, maintaining relatively large area of depositional top surface or weakly erosional platform in the upper part

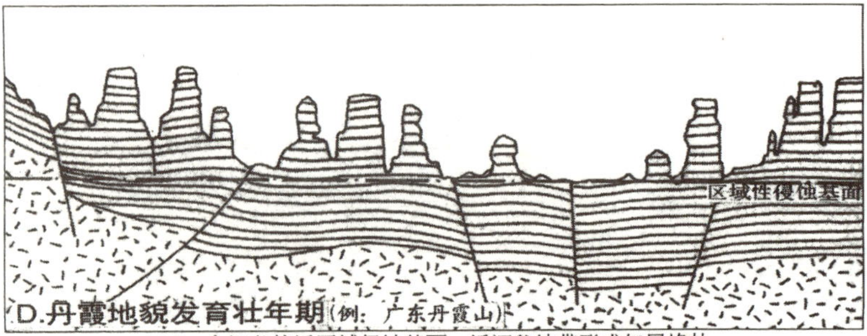

14-D. 主河谷接近区域侵蚀基面，近河谷地带形成红层峰林，
远河谷地带发育红层峰丛，地表最崎岖

The main valley approaching regional erosion basis, red bed hoodoo formed in area close to river and series of peak developed in red bed distant from valley, ground surface most rugged and rough

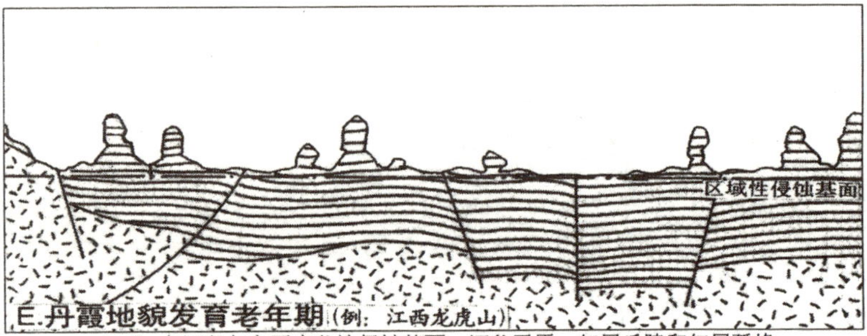

14-E. 主河谷与主要支谷达侵蚀基面，河谷平原、红层丘陵和红层孤峰
相间分布，局部可保持峰林状

The main valley and major branch valleys reaching the erosion basis, distribution of alternating valley plain, red bed hill and red bed isolated peak, with series of peak preserved locally

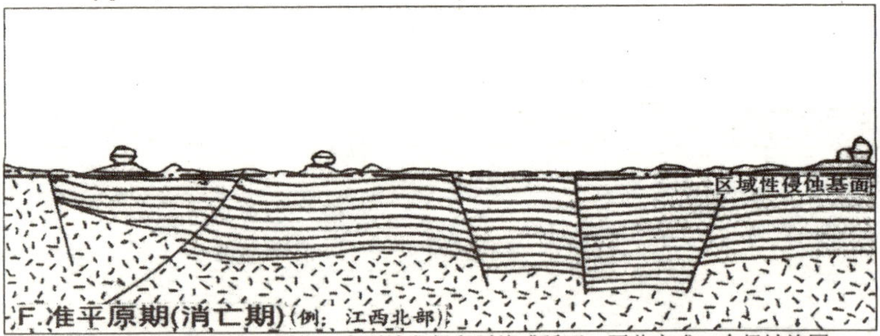

14-F. 波状起伏的准平原化，个别地段保留孤峰或孤石，至此完成一个侵蚀旋回

The region undergoing peneplanation with the ground surface in an undulating form, isolated peak or lonestone preserved locally, then an erosion cycle is completed

图 14　丹霞地貌侵蚀旋回图示

Fig. 14　The erosion cycle of Danxia Landform

(图14 注：假设 B 以后地壳相对稳定。若某阶段地壳抬升，则在此基础上流水下切，形成丹霞地貌的多层性 。Assuming that after stage B, the Earth's crust is relatively stable. If at a certain stage the Earth's crust is uplifted, then the flowing water will cut down, bring about a multi-terrace Danxia landform.)

The development of Danxia Landform begins with the uplifting of red basin. Since red basins are developed in a relatively rigid platform, the late-stage tectonic disturbances are often dominated by fault block activities. Contemporaneous with the uplifting of red basin, major river valley will often develop in an inherited way in the original depression; fractured zone and major joints are the weak zones susceptible to down-cutting by river, and hence may also develop into major river valleys. The top surface of red bed above the local erosion base on either side of the major river valley will first be dissected by flowing water along faults and vertical joints to form deep and narrow dissected valley, which may further deepened and enlarged by erosion into "a thread of sky"-type "tunnel-like valley." In this stage, except for the slopes on either side of the major river valley, collapse is generally not developed on the slopes of the dissected valley or tunnel-like valley.

In the "tunnel-like" branch valley, when the flowing water dissects to a certain depth and encounters the underlying hard stratum or approaches the local erosion base, lateral erosion will take place to destruct the base of the valley walls, which will consequently collapse gradually along vertical joints to make the branch valley widen into a canyon. At the same time, colluvium may be accumulated at the valley bottom, and colluvial gentle slopes will be developed. With the widening of the valley, the colluvial gentle slopes are more distinctly recognizable.

Subsequently, due to the weakening of fluvial erosion, the

base of the escarpment slope is not likely to be attacked directly by flowing water, and the collapse of the escarpment slope will proceed slowly depending on the weathering and caving-in of the incompetent stratum. Huang jin (1982) holds that covered by colluvium, the piedmont bedrock can be protected temporarily. During the collapse and retreat of the escarpment slope, a colluvial gentle slope is formed at the lower part of the piedmont, and a bedrock gentle slope is also formed below the colluvium. With the above-mentioned process going on, the escarpment slope collapses and retreats constantly, and the colluvial slope increases constantly in width and height, while the underlying bedrock gentle slope surface also increases in width and height.

With the collapse and retreat of the valley slope, the gentle slope at hilltop will be dissected and its area diminishes gradually; at the same time the piedmont gentle slope will be enlarged gradually to form a relatively wide range of hills with gentle slopes. The original hill blocks will correspondingly retreat to form "fortress-like residual peaks" or isolated stone columns. With the intermittently uplifting of the Earth's crust or the intermittently dissecting by flowing water, the above-mentioned process may take place repeatedly in an area. Therefore, step-and-platform topography displaying alternating steep and gentle slopes (This occurs also in areas where the profile is characterized by alternating competent and incompetent strata) is often seen in Danxia Landform areas.

In addition, according to Huang Jin, when collapse takes place in escarpment slope, rock blocks falling down will be crushed drastically under the strong action of gravity. This will greatly increase the contact surfaces of the rock fragment and debris with air, water and living things, and thus accelerating the weathering process. Moreover, a large-scale collapse at the escarpment is often followed by a stable period, during which not only sheet-like and block-like weathering and spalling can occur, but also the colluvial deposit on the gentle slope at the lower part of the escarpment can have a stable period to undergo weathering, thus, a considerable part of the gentle slope at the base of the escarpment will turn into red soil and develop a red weathering crust. At the same time, flowing water will wash and erode the gentle slope. When the soil layer on some part of the slope has been eroded off to expose the bedrock underneath, the flowing water will erode the bedrock along joints to form a new escarpment slope (This may also occur when the Earth's crust is uplifted again). Then a new collapse gentle slope overlying a gently dipping bedrock surface will form at the base of the new escarpment, and will enlarge constantly with the collapse and retreat of the escarpment above.

10 种类多样，千变万化，丰富多彩的丹霞地貌类型

Infinite in variety, rich and varied types of Danxia Landform

由于发育丹霞地貌的物质基础、构造条件、外动力因素等差异，形成了丰富多彩的丹霞地貌类型。黄进等(1992)曾以六个方面的依据对丹霞地貌进行不同系列的分类(表 1)。

The difference in material basis, tectonic conditions and exogenic factors etc. has given rise to the rich and varied types of Danxia Landform. Based on six criteria, Huang Jin *et al.* (1992) have classified Danxia Landform into different series (Table 1).

表1 丹霞地貌分类

Table 1 Preliminary classification scheme for Danxia Landform

分类依据 Criterion	类型 Type
岩层倾角 Dip angle of stratum	(<10°)近水平丹霞地貌，(10° ~30°)缓倾斜丹霞地貌，(>30°)陡倾斜丹霞地貌 (<10°) subhorizontal type; (10°~30°) gently dipping type; (>30°) steeply dipping type
有无盖层 Covering stratum	(无盖层)典型丹霞地貌，(有盖层)类丹霞地貌 (Without covering stratum) typical Danxia type; (with covering stratum) Danxia-like type
气候区 Climate	湿润区、半湿润区、半干旱区、干旱区丹霞地貌 Humid, semi-humid, semi-arid or arid region type
发育阶段 Development stage	幼年期、壮年期、老年期丹霞地貌 Young, mature or old age type
有无喀斯特化 Karstification	(有喀斯特化)丹霞喀斯特地貌，(无喀斯特化)非丹霞喀斯特地貌 (With karstification) Danxia karst;(without karstification) non-Danxia karstic landform
地貌形态 Geometry	宫殿式(柱廊状、窗棂状)、方山状、峰丛状、峰林状、石墙状、石堡状、孤峰状等 Palatial type (porticus-shaped, mullion-shaped), mesa-shaped, peak-cluster-shaped, peak-forest-shape, stone-wall-shape, stone-fortress-shaped and isolated-peak-shape

资料来源：根据黄进等有关文献编制。 Data origin: Modified after Huang Jin *et al.*

笔者根据中国丹霞地貌多年研究成果，提出如下分类系统（表2）。

Based on the results of many years' research on Danxia Landform, the writer proposes the following classification system (Table 2).

表2　丹霞地貌分类系统

依据	类　型	指　标	特　　征	分　　布
岩石特征	砾岩或砂砾岩丹霞地貌	厚层、巨厚层砾岩、砂砾岩占2/3以上	垂直剖面岩性变化较大，石质坚硬，陡崖坡高大，凹凸不平，多顺层槽脊，风化坡面粗糙	接近物源的盆地边缘或小型盆地内部
	砂岩丹霞地貌	厚层砂岩占2/3以上	岩性较坚硬，可形成较高大的陡崖坡，坡面变化较小且光平	大沉积盆地的中环或小盆地的内部
	砂页岩丹霞地貌	薄层粉砂岩、泥质岩占2/3以上	岩性较软，陡崖坡不高、不平整、不连续且坡度较缓，坡角风化碎屑较多	大盆地中部或高差不大的小盆地
岩层倾角	近水平丹霞地貌	岩层倾角小于10°	山顶形态与层面接近，呈微上凸形，四壁陡立，方山状丹霞地貌	构造平缓区或盆地中部
	缓倾斜丹霞地貌	岩层倾角介于10°~30°之间	山顶面与单斜层面接近，除顺向坡外，其他三面多为陡崖，呈单面山状丹霞地貌	断层线附近或盆地边缘
	陡倾斜丹霞地貌	岩层倾角大于30°	若其山顶没有古夷平面，则发育尖顶状丹霞地貌，逆向坡的陡崖较为发育	构造变动强烈区或盆地边缘

续表 2

依据	类型		指标	特征	分布
单体地貌形态	正地貌	丹霞崖壁	坡度>60°，高度>10 m 的陡崖坡	多为直立陡崖，可因岩性差异呈多层状组合；壁上多顺层凹凸和竖向流水蚀槽	坚硬岩石的绝大部分边坡或河流侵蚀部位
		丹霞方山	近平顶,四面陡坡,长宽比小于2:1	岩层近水平，山顶平缓，四壁陡立，呈城堡状、宫殿式丹霞地貌	近水平岩层分布区，构造盆地中部
		丹霞石墙	长度大于2倍宽度,高度大于宽度	山块顺断裂构造线延伸，呈薄墙状，低缓者可称石梁	近水平岩层分布区，构造盆地中部，或垂直断裂切割成条块状的地带
		丹霞石柱	孤立石柱,高度大于直径	方形或圆形孤立石柱，低矮者可称石墩	近水平岩层分布区，构造盆地中部，或垂直断裂切割成方块的地段
		丹霞尖峰	由陡崖坡构成的锥状山峰	四面陡坡，局部有陡崖，但山顶面不发育，呈锥状山峰	无明显规律
		丹霞低山	局部陡崖,多为30°~60°的陡坡山峰	可能有1~2个面呈局部陡崖坡，大部以陡坡或陡缓坡相间构成山峰或山梁	无明显规律或岩性稍软地段
		丹霞丘陵	局部有陡崖,山顶浑圆化的低缓丘陵	无连续陡崖坡，总体上呈圆化丘陵状	老年期的丹霞地貌或软岩地段

续表 2

依据	类型		指标	特征	分布
单体地貌形态	正地貌	丹霞石球	浑圆状风化、蚀余球状石	由坚硬的厚层、巨厚层砂岩或砂砾岩风化或侵蚀而残余的球状石	老年期残余,分布无明显规律
		崩积锥和崩积巨石	陡崖下不规则锥状崩积体和巨石块	块状崩塌堆积，叠置洞穴；石块大小不同，单块巨石大可至几百立米	陡崖坡的下部
	负地貌	沟谷	同一般沟谷	主河谷多宽谷；支谷多峡谷；源头多巷谷；落差较大的谷底多壶穴	主要顺构造破碎带发育,或继承原始洼地
		顺层凹槽	顺软岩层发育的凹槽，深度小于槽口高度	岩性垂向差异使崖壁上出露的软岩层快速风化成凹槽，顺层可连续或不连续	崖壁上软岩出露地带
		丹霞洞穴	深度大于外口最小尺度(高或宽)的凹穴	顺层延伸较长(宽)如额状洞，扁平洞；纵深方向多顺层延伸	崖壁软岩带或流水侧蚀部位
		丹霞穿洞	蚀穿山块的通透洞穴	通透洞穴即穿洞(穿岩,石窗)；洞顶厚度小于跨度者为石拱；拱跨在河谷者称天生桥	丹霞石墙中软岩地段，流水侧蚀部位
		竖向洞穴	高度大于宽度,垂直方向延伸的洞穴	顺垂直裂隙发育的垂向洞，或崖壁表面水流侵蚀的竖向洞穴	大节理交叉部位的下水通道

续表 2

依据	类　型	指　标	特　征	分　布
按气候区分	湿润区丹霞地貌	按气候区标准	山顶和缓坡植被覆盖度好，流水作用面较光滑，陡崖坡上多竖向流水侵蚀浅槽	湿润区
	半湿润区丹霞地貌	按气候区标准	山顶和沟谷植被覆盖较好，表面物理风化较强，地貌特点接近湿润区	半湿润区
	半干旱区丹霞地貌	按气候区标准	植被较少，温差风化和盐风化作用较强，有一定的风蚀作用，坡面较粗糙，片状剥落较明显	半干旱区
	干旱区丹霞地貌	按气候区标准	基本无植被，温差风化和盐风化强烈，风蚀作用使表面形成壁龛，表面层状剥落明显，坡面粗糙	干旱区
	高寒区丹霞地貌	按气候区标准	基本无植被，寒冻和温差等物理风化强烈，层状剥落明显，裸岩表面粗糙	高原高寒区

续表 2

依据	类型	指标	特征	分布
按组合结构分	高原峡谷状丹霞地貌区	保持大面积高原面，沟谷深切	山顶面基本保持统一高原面(古剥夷面)，无或少有孤立山块，丹崖集中在峡谷两坡	近期地壳强烈块状抬升区
	山岭状丹霞地貌	以起伏较大的锥状山峰为主的红层山地	岩性相对软弱的红层，在高原峡谷型的基础上进一步侵蚀形成的红层山地	地壳抬升，抬升后侵蚀时间较长
	峰丛状丹霞地貌	以直立形态为主，但基座相连的丹霞地貌区	岩性坚硬，上部形成直立丹霞地貌，下部尚未切透或地壳再度抬升，基座高度大于上部	地壳抬升，抬升后有较长的稳定期
	峰林状丹霞地貌	基座分离，以直立形态为主的丹霞地貌区	岩性坚硬，山块分离，柱状墙状方山状山块林立，沟谷剖面较平缓	地壳抬升，抬升后有较长的稳定期
	孤峰状丹霞地貌	山峰孤立，或山峰间距大于山峰高度	丹霞岩峰呈孤立状，不成山区；或峰与峰疏散分布，之间有宽阔的平缓谷地或缓丘	抬升后长期稳定的侵蚀区
	丘陵状丹霞地貌	山峰浑圆化，大部分山坡平缓的丹霞地貌区	无丹霞石峰，山顶浑圆，局部保持陡崖，丘顶或谷地偶见丹霞石球	抬升后长期稳定的侵蚀区

续表 2

依据	类型	指标	特征	分布
按发育阶段	幼年期丹霞地貌	山顶保持古剥夷面，以峡谷巷谷为特色	同高原峡谷型丹霞地貌	堆积后或侵蚀后的强烈抬升侵蚀区
	青年期丹霞地貌	山顶分离，但与古剥夷面相当，支谷仍为峡谷	高原峡谷型丹霞地貌进一步切割,山顶呈山原面，除主河谷外峡谷和巷谷仍为负地貌主体	抬升后的侵蚀区
	壮年期丹霞地貌	峰丛状，峰林状特征突出	同峰丛状、 峰林状丹霞地貌	抬升后的长期稳定侵蚀区
	老年期丹霞地貌	孤峰状或丘陵状丹霞地貌	同孤峰状、丘陵状丹霞地貌	长期稳定侵蚀区
过渡与组合类型	喀斯特化丹霞地貌	有明显的化学侵蚀和化学沉积作用	依 $CaCO_3$ 含量<30%为弱喀斯特化丹霞地貌，>30%为强喀斯特化丹霞地貌	物源地石灰岩较多的地区
	泥被式丹霞地貌	丹霞地貌有上部泥浆在表面涂盖	丹霞地貌被黄土覆盖,表层水流携带泥浆在其表面沉积，崖壁上形成垂向的泥条	黄土分布区
	土林式丹霞地貌	偏红色土状堆积物发育的地貌	成岩作用较差的土状堆积,受侵蚀后形成的类似地貌，表面粗糙，土柱、土墙、土堡等	新第三纪或第四纪早期堆积而现代抬升区

Table 2 Classification system for Danxia Landform

Criteria	Type	Index	Characteristics	Distribution
Litho-logies	Conglomerate or sandstone	More than 2/3 are thickly bedded or very thickly bedded conglomerate or sandstone	Great vertical variation in lithologies, the rock is hard, the escarpment slope large and high, uneven, bedding grooves abundant, weathered slope surface rough	At the basin margin or within small basin close to provenance
	Sandstone	More than 2/3 are thickly bedded sandstone	Rock relatively hard and can form high and large escarpment slope; the slope surface is smooth, straight and rarely variable	At middle ring of major sedimentary basin or within minor sedimentary basin
	Sandshale	More than 2/3 are thinly bedded siltstone and pelitic rocks	Rock relatively soft; escarpment slope is not high, not even, not continuous, and is relatively gentle, with abundant weathered detritus at foot	At middle part of major basin or minor basin with small drop in level

Continue Table 2

Criteria	Type	Index	Characteristics	Distribution
Dip angle of strata	Subhorizontal	Dip angle of strata less than 10°	The hilltop surface follows the bedding plane in geometry, slightly convex upward; faces around the hill rise steeply; mesa-shaped Danxia landform	In gentle structure area or in the middle part of a basin
	Gently dipping	Dip angle of strata ranges between 10° and 30°	The hilltop surface follows the monoclinal bedding plane; except for the consequent slope, the other three sides are mostly escarpments, occurring as a cuesta-shaped Danxia landform	At the vicinity of fault line or basin margin
	Steeply dipping	Dip angle of strata greater than 30°	In case there is no palaeo-planation surface at hilltop, pointed-top-shaped Danxia landform will develop, and escarpment along obsequent slope is well developed	In violent structure athletics area or basin margin

Continue Table 2

Criteria	Type		Index	Characteristics	Distribution
Single landform geometry	Positive relief	Danxia escarp-ment face	Escarpment slope with a gradient >60° and a height >10 m	Mostly upright escarpment; multi-terrace association can be formed due to difference in lithologies; protrusions and depressions along beddings and vertical flowing water eroded grooves are common on the face	On most of the slopes formed by hard rock or in those parts eroded by river
		Danxia mesa	Nearly flat-top with steep slopes at the margin; length-width ratio <2:1	Strata nearly horizontal; hilltop flat-lying, with upright faces at the margin, displaying a castle-shaped or palatial type Danxia landform	In areas where subhorizontal strata are distributed and in middle part of structural basin
		Danxia stone wall	Length greater than twice of the width; height>width	Hill mass extending along fracture lines, occurring as thin wall; those low-lying may refer to as stone beams	Same as above, or in areas where the rock mass has been cut perpen-dicular to fractures into stripes
		Danxia stone column	Isolated stone column; height> Diameter	Isolated stone column, square or round shaped on plan view; those low-lying may refer to as stone blocks	Same as above, or in areas where the rock mass has been cut perpen-dicular to fractures into square sections
		Danxia spire	Taper peak composed of escarpment slopes	With steep slopes and locally escarpment on all sides; top surface not developed, occurring as taper peak	No notable distribution regularity

Continue Table 2

Criteria	Type		Index	Characteristics	Distribution
Single landform geometry	Positive relief	Danxia low hill	Locally escarpment, mostly peaks with 30°-60° steep slopes	May have 1-2 faces occurring as escarpment slopes, mostly peaks or flat-topped ridges made up of steep slopes or alternating steep and gentle slopes	No notable distribution regularity, or in areas where the rock is slightly soft
		Danxia hill	Low-lying hill with rounded top and locally with escarpment.	No continuous escarpment, with an overall appearance characterized by a rounded hill	In soft rock terrains or in areas where the Danxia landform has evolved into old stage
		Danxia stone ball	Residual ball-shaped stone formed by onion weathering	Residual ball-shaped stone formed by onion weathering and erosion of hard, thickly or very thickly bedded sandstone and conglomerate	Residue in old stage; no notable distribution regularity
		Colluvial cone and huge colluvial rock block	Irregular cone-shaped colluvial mass and huge rock blocks at the foot of an escarpment	Block-like colluvial accumulation, superimposed cave; rock blocks are of varied sizes, and a block may be as huge as several hundred or a thousand cubic meters in volume	At the food of an escarpment

Continue Table 2

Criteria	Type		Index	Characteristics	Distribution
Single landform geometry	Negative relief	Valley	Same as the average valley	Major valleys are mostly wide valleys, while branch valleys, canyons; tunnel-like valleys are common in headstreams, while pot-holes, in valley bottoms with great fall head	Developed mainly along structural fractured zones, or inherited from primary depressions
		Bedding notch	Notches developed along beddings of soft strata have their depths less than the heights of the notch mouths	Lithologic difference in vertical direction accelerates the weathering of soft stratum exposed on escarpment face, resulting in the formation of notches, which may be continuous or discontinuous along beddings	At such portions of the escarpment face where soft rock is exposed
		Danxia cave	Alveolization with its depth greater than the smallest dimen-sion of its outer mouth	Extending rather long (wide) along bedding, such as forehead-shaped cave, flat cave; the depth direction is mostly parallel to the bedding.	Soft rock portion or flowing water laterally eroded portion of the escarpment face
		Danxia perforated rock	Through-holes formed by erosion through the entire hill block	Through-holes are also known as perforated rock (stone window); those with hole top thicknesses less than span lengths are known as stone arch; while their arches stride across river valleys are called natural bridge	Soft rock portion or flowing water laterally eroded portion of Danxia stony wall
		Vertical cave	Caves extending vertically with their heights greater than their widths	Vertical caves developed along vertical fractures, or formed by flowing water erosion on escarpment faces	Drainage at intersection of major joints

Continue Table 2

Criteria	Type	Index	Characteristics	Distribution
Climate	Danxia landform in humid region	Based on climatic region classification	Hilltop and flat slope where vegetation coverage is high; the surface on which flowing water acts is relatively smooth; flowing water eroded vertical shallow grooves are abundant on the escarpment face	Humid regions
	Danxia landform in semi-humid region	- ditto -	Hilltop and flat valley where vegetation coverage is high; physical weathering is stronger at rock surface	Semi-humid regions
	Danxia landform in semi-arid region	- ditto -	Vegetation is rare, weathering due to temperature variation and salinization is strong, while wind erosion is more or less active; the slope surface is rough	Semi-arid region
	Danxia landform in arid region	- ditto -	Basically without vegetation, weathering due to temperature variation and salinization is very strong, while wind erosion give rise to stable holes on the surface; the slope surface is rough	Arid region
	Danxia landform in subnival region	- ditto -	Basically without vegetation, physical weathering due to freezing and temperature variation is very strong, stratified spalling is significant, bare rock surface is rough	Subnival region in plateau

Continue Table 2

Criteria	Type	Index	Characteristics	Distribution
Combination structure	Plateau-canyon-shaped Danxia Landform	Large area of plateau surface is preserved with deeply dissected valleys	Top surfaces of hills are the original plateau surface (palaeoplanation surface); without or with only a few isolated hill mass; red escarpments occur on either side of the canyon	Recent intensely block-like uplifted area of the Earth's crust
	Mountain-ridge-shape Danxia Landform	Red bed mountain area dominated by conical peaks of high relief	Red bed mountain area formed in soft red beds through further erosion of the plateau-canyon-shaped Danxia landform	Same as above, but the erosion period after uplift is relatively long
	Peak-group-shaped Danxia Landform	Danxia Landform area dominated by upright-shaped peaks with their bases connected together	The rocks are hard, upright-type Danxia landform is formed in the upper part, while the lower part has not been cut through though the Earth's crust is uplifted again; the lower part is higher than the upper part	Same as above, but the stable period after uplift is relatively long

Continue Table 2

Criteria	Type	Index	Characteristics	Distribution
Combination structure	Hoodoo-shaped Danxia Landform	Danxia Landform area dominated by upright-shaped peaks but with their bases separated from each other	The rocks are hard, the hill masses are separated from each other, column-shaped, wall-shaped and mesa-shaped hill masses stand in great numbers, the valley bottoms are relatively flat	Same as above
	Isolated-peak-shaped Danxia Landform	Peaks are isolated from each other; the spacing between peaks is greater than the heights of the peaks	Danxia-type peak stands in isolation in a non-hilly area; or peaks are sparse and scattered, with open and flat valley land or low hills in between	Long-term erosion area
	Hill-shaped Danxia Landform	Danxia Landform area where the peaks are rounded, and most of the slopes are gently dipping	No Danxia-type peak is present; the hilltop is rounded, escarpment is preserved locally, while stone balls are occasionally seen on hilltop or on valley land	Long-term erosion area

Continue Table. 2

Criteria	Type	Index	Characteristics	Distribution
Development stage	Juvenile-stage Danxia Landform	Trace of palaeoplanation is preserved on hill top, characterized by tunnel-like valley and canyon	Same as the Plateau-canyon-shaped Danxia Landform	Intense uplift-erosion area after accumulation or erosion
	Young-stage Danxia Landform	Hilltops are separated from each other, but their horizons correspond to that of the palaeoplanation, the branch valleys are still of canyon type	The plateau-canyon type Danxia Landform is further dissected; the hill top is subrounded; except for the major valley, the canyon and the tunnel-type valley are still dominating the negation landform	Erosion area after uplifting
	Mature-stage Danxia Landform	Significantly characterized by peak-group type and hoodoo type Danxia Landform	Same as the peak-group-shaped and hoodoo-shaped Danxia Landform	Long-term stable and erosion area after uplifting
	Old-stage Danxia Landform	Isolated-peak-shaped and Hill-shaped Danxia Landform	Same as Isolated-peak-shaped and Hill-shaped Danxia Landform	Long-term stable and erosion area

Criteria	Type	Index	Characteristics	Distribution
Transitional and composite Danxia landform	Karstified Danxia Landform	With significant chemical erosion and chemical deposition	Subdivided into weakly karstified ($CaCO_3$ <30%) and strongly karstified ($CaCO_3$>30%) Danxia Landform	Area where the provenance contains abundant limestone
	Mud-coating type Danxia Land form	Danxia Landform coated by mud	Danxia Landform covered by loess; surface flowing water carrying mud to deposit on the landform; hanging mud strings are seen on escarpment face	In loess terrain
	Earthy forest-type Danxia Landform	Landform developed in slightly red earthy accumulation area	Landform of this kind is formed by erosion of earthy accumulations which has undergone uncompleted diagenesis; the surface is rough; earthy column, earthy wall and earthy fortress are common	In Late Tertiary or Early Quaternary accumulation area that experienced recent uplifting

11 在中国
丹霞地貌分布广泛

In China
Danxia Landform is widely distributed

中国丹霞地貌分布广泛，根据近年国内各地学者的研究，在亚热带湿润区，温带湿润区、半湿润区、半干旱和干旱区，青藏高原高寒区都有分布，目前已知共有503 处（据黄进统计）。我国地理环境的区域差异使丹霞地貌的发育特征表现出一定的差异性。不同的气候带产生的外力组合，以及晚近地质时期环境的变迁，都不同程度地影响丹霞地貌的发育进程和地貌特征的继承与演变。至1999年5月按黄进总结的503处丹霞地貌分布状况以省区归纳如表3。

Danxia Landform is widely distributed in China. According to the results of recent years from researchers of various places, altogether 503 Danxia Landform locations (according to Huang Jin) are known in subtropical humid regions, temperate humid regions, semihumid regions, semiarid regions, arid regions and subnival regions in Qinghai-Tibet Plateau. Because of the regional difference of geographic environment in China, Danxia Landform is different to a certain degree from place to place. The exogenic force association prevailing in different climatic zones, and the environmental variation that took place in recent geologic times will both affect to different degrees the development process of Danxia Landform as well as the inheritance and evolution of its features. According to Huang Jin's statistics, up to May 1999 the 503 known locations of Danxia Landform distributed in the various provinces, cities and regions are shown in Table 3.

表3 中国各省、市、自治区丹霞地貌一览表

Table 3 Locations of Danxia landform in various provinces, cities and autonomous regions in China

序号 No.	所在省、市、自治区 Province/city/region	处数 Quantity	序号 No.	所在省、市、自治区 Province/city/region	处数 Quantity	序号 No.	所在省、市、自治区 Province/city/region	处数 Quantity
1	广东省 Guangdong	46	10	重庆市 Chongqing	10	19	宁夏回族自治区 Ningxia	2
2	江西省 Jiangxi	72	11	湖北省 Hubei	7	20	海南省 Hainan	1
3	福建省 Fujian	32	12	陕西省 Shanxi	11	21	甘肃省 Gansu	65
4	浙江省 Zhejiang	37	13	河南省 He'nan	2	22	青海省 Qinghai	32
5	安徽省 Anhui	1	14	河北省 Hebei	2	23	云南省 Yunnan	10
6	湖南省 Hu'nan	32	15	辽宁省 Liaoning	2	24	江苏省 Jiangsu	4
7	广西壮族自治区 Guangxi	14	16	黑龙江省 Heilongjiang	1	25	内蒙古自治区 Neimonggu	1
8	贵州省 Guizhou	22	17	山东省 Shandong	1	26	台湾省 Taiwan	1
9	四川省 Sichuan	85	18	新疆维吾尔自治区 Xinjiang	10		合计 Total	503

资料来源：根据黄进的资料统计，数据截止到1999年5月。Data origin: Statistic on the data from Huang Jin, close may 1999.

12 丹霞地貌：一种宝贵的旅游资源

Danxia Landform: A valuable tourism resource

丹霞地貌分布区内往往沟壑纵横，起伏剧烈，风化壳较薄，土壤养分较少，生态平衡脆弱，不宜作为工、农、交通用地。但典型的丹霞地貌山块离散，群峰成林；赤壁丹崖上色彩斑斓，洞穴累累；高峡幽谷，清静深邃；石堡、石墙、石柱、石桥造型丰富，变化万千，其雄险可比花岗岩大山，奇秀不让喀斯特峰林；红层盆地中又多有河溪流过，丹山碧水相辉映。因此，丹霞地貌是构成风景名山的一支重要地貌类型（见附图片）。在中国119处国家级风景名胜区中，就有丹霞山、武夷山、龙虎山等22处名山全部或部分由丹霞地貌构成，另有几十处省级风景区和一批具备国家级风景区条件的丹霞地貌名山。作为一种独特的风景地貌用于旅游开发，比较其他功能的开发更能实现资源价值，更有利于兼顾经济、社会和环境效益。随着人们的旅游需求不断增加，其旅游价值将得到更加充分显示。

改革开放以来，我国旅游业发展迅速，众多的丹霞地貌风景区逐步开发，对丹霞地貌的基础研究和旅游开发实践研究提出了新的要求。1991年成立丹霞地貌旅游开发研究会以来，各地学者积极参与研究区域丹霞地貌旅游资源评价与开发实践工作，根据不同需要从相关学科引进了先进的研究方法和思维理念，使丹霞风景地貌的研究逐步走上规范化。

In areas of Danxia Landform, ravines and gullies criss-cross, bringing about very rugged relief. The weathering crust is thin, the soil is lean, the ecological environment is not rich, and no important mineral resources have ever been discovered. Therefore it is not advisable to use the land for industrial, agricultural or communication purposes. However, the hill masses of typical Danxia Landform are scattered about, peaks are grouped into forest; the red walls and deep red escarpments are beautifully colored, with caves occurring in clusters; the gorges are deeply cut and the valleys are deep and secluded, therefore the places are very tranquil; stony fortresses, stony walls, stony columns and stony bridges are of varied shapes and most changeful. Such areas are comparable to granitic mountain for their magnificent appearance and precipitousness, and to karst hoodoo for their wonders and beautiful views; besides, most of the red basins have rivers and stream flowing through, and red hills and green water add radiance and beauty to each other. Therefore Danxia Landform is one of the important types of landforms that form the famous landscapes and well-known mountains of China. Among the 119 State-grade scenic spots (Chinese national park), more than 20 are formed by Danxia Landform, including, for example, Mt. Danxiashan, Wuyishan and Longhushan. In addition, there are several tens of provincial-grade scenic spots and a number of scenic spots that have the prerequisites to become State-grade (national park) are made up of Danxia Landform. These resources are indeed more valuable to treat as peculiar landscapes used for tourism development than to use for other purposes. Besides, to use it for tourism development is more favorable for the implementation of the policy of "taking into consideration of not only the economic, but also the society and the environmental benefits." With the ever increasing tourists, the value of Danxia Landform as a scenic resource will become more and more apparent.

Since the implementation of the "Reform and Open up" policy, China's tourist industry has been developing very rapidly. Many scenic spots of Danxia Landform have gradually been developed. This has set on new demands on the basic research of Danxia Landform and on the tourism development practice. Since 1991 when The Danxia Landform and Tourism Development Research Society of China was set up, researchers from various places have actively taken part in regional Danxia Landform research programs concerning the evaluation of tourism resources and the improvement of development practice. To satisfy the needs in various aspects, advanced research methods and new ideas have been introduced from the relevant disciplines, and the research procedures of Danxia Landform will be normalized step by step.

13 特殊的地貌，特殊的文化载体

Special landform, special cultural carrier

彭华(1996)以丹霞山为例研究了丹霞山水文化的旅游开发，认为丹霞山水是一个“山水景观综合体”，山水文化不仅是人类在名山中创造的文化，还涉及到关于丹霞地貌的地理学、地质学、地貌学、水文学、生物学、气象与气候学、山水美学和“附会文化”等知识。这些科学与美学知识，只要按照旅游需求加以策划和组织，就可转化为文化旅游产品。但这方面的开发目前只限于学生的科普教育和教学实习，而面对大众的科普旅游仍然是一个弱项。从这个意义上讲，丹霞山水文化是一项大有潜力可挖的文化旅游资源。

黄进(1992)、韩振飞(1992)、张忠孝(1998)等曾研究了人文景观与丹霞地貌的配置关系，除了作为特殊的自然旅游资源外，丹霞地貌区还孕育和沉淀了丰富的人类文化遗产。

丹霞地貌最突出的形态要素是雄伟的赤壁丹崖，单体或群体形态上常呈完整的块状或城堡状，紫红色调，给人庄重和神圣之感，同我国传统表现权威、富贵、吉祥的色调一致，也是中国宗教崇尚的主色调，从环境角度加强了宗教场所的威严感和神秘感。这种自然风光与神秘意境的结合，常使丹霞地貌区成为宗教圣地。如集儒、释、道三教于一山的福建武夷山，道教名山江西龙虎山、安徽齐云山、四川青城山、甘肃崆峒山等均由丹霞地貌构成。青海坎布拉则以藏传佛教复兴期发祥地而闻名于国内藏区和海外。

组成赤壁丹崖的厚层红色砂岩层，其岩性结构致密均一，硬度较小，易于精雕细刻，因而留下大量摩崖石刻、摩崖造像、崖壁画等。如四川大足县有“石刻之乡”的美称，乐山大佛山体造像享誉海内外；而赤壁上遍布的浅平丹霞岩洞往往成为石窟艺术的载体，如甘肃麦积山石窟、炳灵寺石窟等石窟艺术成为甘肃最富有特色的旅游资源，因而有“甘肃无处不石窟，石窟无处不丹霞”之说；赣州通天岩的摩崖造像，齐云山、武夷山的摩崖石刻等都是历史和宗教文化珍品。

丹霞地貌区以地形险要而著称，古代的军事要塞、古山寨常凭险而筑，如四川剑门关、广东金鸡岭险关、丹霞山遍布山顶的古山寨等。此外，丹霞赤壁上众多洞穴为居住、悬棺墓、文物存放等提供了天然场所；特别是龙虎山、武夷山绝壁上大量春秋战国时期的崖墓群，存放在距地面上百米的丹霞洞穴内，成为墓葬奇观。

Taking Mt. Danxia as an example, Peng Hua has studied the hills-and-waters culture and tourism development aspects of Danxia Landform. He believes that Danxia hills and waters are a "hills-and-water landscape complex," and that the hills-and-waters culture involves Danxia Landform geography, geology, geomorphology, hydrology, biology, meteorology, climatology, hills-and-waters aesthetics, "analogical culture" and other branches of science. All these branches of science and aesthetic knowledge concerned with Danxia Landform (which can be referred to as Danxia hills-and-water culture) can be organized to meet the needs of tourism and to serve for tourism. However, at present this hills-and-water culture is only a course in universities, and not used to serve for touring and sightseeing, or touring for the purpose of educational activities in popularization of sciences. In this sense, Danxia hills-and-waters culture is a tourist resource that has great potentialities to be exploited.

Huang Jin (1992), Han Zhenfei (1992), Zhang Zhongxiao (1998) and others have studied the relationship between humane landscape and Danxia Landform in an overall layout. Besides serving as a special natural resource for tourism, Danxia Landform areas are also spots rich in human cultural heritage.

The most outstanding geometric element featuring Danxia Landform is the magnificent red walls and deep red escarpments. Geometrically, these, singly or in group, often occur in perfect block form or castle form, tinted with purplish red color, giving people an impression of solemnity and holiness. This purplish red color is in conformity with Chinese tradition to represent authority, riches, honor, and good luck, and is also a major color upheld by Chinese religious circle because it helps to strengthen the solemn and mysterious feeling of a religious place from an environmental angle.

Because of this combination of natural landscape with mysterious atmosphere, Danxia Landform areas are generally the places of the Holy Land. For example, Wuyishan of Fujian, which incorporates Confucianism, Buddhism and Taoism in a mountain, have Danxia Landform, and so have Longhushan of Jiangxi, Qiyunshan of Anhui, Qingchengshan of Sichuan and Kongtongshan of Gansu, all of which are famous mountains of Taoism. Another Holdy Land displaying Danxia Landform is Kanbula in Qinghai, which is well known abroad and among the Zang people as the birthplace of Lamaism in the rejuvenation age.

The thickly bedded red sandstone constituting red walls and deep red escarpments is dense and homogeneous in texture and not very hard, hence it is not difficult to carve fine sculpture out of it. As a result, many inscriptions and bas-relief on escarpments and many frescoes are preserved. For example, the Dazu County of Sichuan has always enjoyed the reputation of being a "Home of Stone Inscriptions," and the "Great Stone Buddha in Leshan Mountain" engraved out of the mountain mass has gained a high reputation at home and abroad. Meanwhile, grotto art is often noted in those shallow and flat Danxia caves distributed throughout some red escarpments. For example, the grotto art in Maijishan grotto and in Bingling Temple grotto has become the most characteristic resources for tourism in Gansu. Therefore, as the saying goes, "There is no place without grotto, and there is no grotto without Danxia in Gansu." In addition, the bas-relief on escarpment in Tongtianyan of Ganzhou, and the inscriptions on escarpments in Qiyunshan and Wuyishan are all historic, religious and cultural treasures.

Danxia Landform areas are celebrated for their strategical locations and difficulty of access, resorting to which were often built ancient military forts and ancient mountain fastnesses, such as the Jianmenguan Pass in Sichuan, the Jinjiling strategic pass in Guangdong and the ancient mountain fastnesses distributed throughout the hilltop of Mt. Danxia. Besides, the numerous caves on the Danxia red escarpments have provided natural places for dwelling, and for hanging coffin and cultural relic to place in. In particular in Longhushan and Wuyishan, many groups of rock-tombs were build in the Spring and Autumn Period (770~476 B.C.) and Warring States Period (475~221 B.C.) in Danxia caves in escarpments several hundred meters above the ground, which is indeed a wonder concerning graves.

14 研究丹霞地貌，为了更好地利用丹霞地貌

To study Danxia landform for using it better

丹霞地貌旅游开发研究会成立以来，在黄进主持下组织召开了五届丹霞地貌旅游开发学术讨论会，除了推动中国丹霞地貌的研究之外，都为会议所在地的旅游开发做出大量的研究工作，提出不少好的建议，同时提高了当地的知名度，促进了丹霞地貌区的旅游经济发展。近年来，丹霞地貌旅游开发研究会的成员均不同程度地介入了丹霞地貌区域的旅游开发实践研究，通过写文章、提意见、做规划等形式为地方献计献策，向有关部门和领导推荐，向社会作宣传。陈传康、彭华等较先介入丹霞地貌旅游区的评价与开发研究，1988年成功地完成了丹霞山风景名胜区基础研究和国家级风景名胜区的报批，并于1989年完成风景区总体规划。彭华1992~1995年曾

直接主持丹霞山阳元石景区和翔龙湖景区的规划、设计与开发实施。黄进、彭华等1994年为丹霞山申报国家地质地貌自然保护区编制的全套上报材料，得到国家自然保护区评委会的全票通过，1995年获国务院批准； 1997年彭华又完成了丹霞山国家地质地貌自然保护区总体规划的编制。

Since the establishment of The Danxia Landform and Tourism Development Research Society of China, five symposia on Danxia Landform and tourism development have been convened with Huang Jin as presider and organizer. This not only has given impetus to the research work on Danxia Landform in China, but also has promoted the tourist economic development of the Danxia Landform areas. Besides, a large amount of research work were done before symposia and many good proposals were put forward during symposia concerning the tourism development of the places where the symposia were located, while at the same time the notability of these places has also been raised. In recent years, members of the Research Society have contributed to different degrees to the study on tourism development and practice for the Danxia Landform region. Through publishing papers, putting forward proposals and working out development schemes, they offer advice to the local authorities, give recommendations to leaders of the relevant departments, and spread propaganda in the society. Chen Chuankang, Peng Hua and others are among the earliest to contribute to the evaluation and development study of the Danxia Landform tourist spots. In 1988, they completed the basic research project on Mt. Danxia scenic spot and succeeded in applying for approval for classifying Mt. Danxia as a stategrade scenic spot. In 1989, they completed the overall planning of this scenic spot. In 1992~1995, Peng Hua was in charge directly of the planning, design and development of the Yangyuanshi and the Xianglong Lake scenic spots of Mt. Danxia. In 1994, Huang Jin, Peng Hua and others compiled a complete set of materials concerning classifying Mt. Danxia as a state geological-geomorphologic natural preservation area. These set of materials was reviewed and passed by a unanimous vote in a meeting held by the Reviewing Committee on State Natural Preservation Areas, and was approved by the State Council in 1995. In 1997, Peng Hua completed compilation of the overall planning for Mt. Danxia State Geological-Geomorphologic Natural Preservation Area.

15 未来：美好的期盼

The future: Magnificent prospects

15.1 丹霞地貌研究的理论与实践意义

丹霞地貌是地壳演化到一定历史阶段而出现的特殊地貌类型，其地层、岩性、沉积环境、构造、坡面发育、形态特征都表明它的特殊性。丹霞地貌及其相应的基础研究将有助于揭示中生代以来的地壳演化特征，深化和丰富地貌学理论。丹霞地貌研究是在中国诞生、发展，并逐步走向成熟的一个地貌学分支学科，是中国地学工作者对科学的一份贡献。

丹霞地貌在中国有着广泛的分布，其类型多样，发育典型。研究丹霞地貌区域的资源保护、环境培育和开发利用方向，对于区域社会经济发展有着巨大的实践意义。尤其是作为一种优良的风景地貌，具有很高的旅游开发价值。丹霞地貌风景资源转化为经济、社会效益的工作日益受到社会各界的重视。因此，丹霞地貌旅游开发研究工作者面对着为生产建设服务的、意义重大的研究任务。

15.2 加强基础理论研究

近年来丹霞地貌的有关基础研究虽然取得了不少的成果，但与岩石地貌学的其他学科尤其是与喀斯特地貌相比，丹霞地貌的理论建设仍显得比较薄弱，尚未形成完整的理论体系。今后应更好地利用岩石学、构造地质学、普通地貌学等学科的成熟理论，广泛吸收现代研究方法，推动学科深入发展，在丹霞地貌的形成条件、动力机制、演化过程与阶段、分类系统等方面作进一步的深入研究，建立丹霞地貌的完整理论体系。

15.3 扩大应用研究的接触面

旅游开发和旅游经济发展的需要，对丹霞地貌应用研究将提出更高的要求，尤其是对于科学的评价系统、市场分析和旅游经济发展、开发策划与规划、科普文化提炼等，将成为研究者必须面对的新课题。预计今后具有更强操作性的应用研究将占更重要的地位。

目前的旅游开发应用研究大多停留在半成品层次，即景观描述+开发建议，具有很大的随意性。我们提倡的可操作性可分为三个层次：宏观层次要求在战略方向上为政府决策提供依据，中观层次要求能够制定开发布局与资源保护规划，微观层次要求在小区规划与设计、项目论证与建设上能够具体实施。因此，应加强与应用地貌学、旅游地理学、旅游资源调查和评价、旅游开发原理和旅游经济学等学科的结合，扩大应用研究接触面，开创新的生长点，更直接、更深入、更有效地介入旅游开发实践中，以增加学科发展机会。

15.4 尽快与国际研究接轨

从目前了解的情况看，世界上主要古陆台都有一定的红层分布，其中已知美国西部、澳大利亚、法国、巴西等地有发育良好的丹霞地貌。但国际上还没有将其作为一个独立的地貌类型进行研究。我国丹霞地貌研究已经取得了大量成果，与国际交流与合作研究的条件已经成熟。近期内应通过地质、地理界已经形成的各种渠道，争取国外学者参加研究与交流，介绍国外的丹霞地貌及研究进展，在适当的时候组织召开国际学术讨论会。通过这一工作，使丹霞地貌形成更广泛的国际影响，实现与国际研究接轨，最终确立丹霞地貌在国际地貌学分类系统中的地位。

(1) Theoretical and practical significance in the study of Danxia Geomorphology: Danxia Landform is a peculiar type of landform formed at the time when the Earth's crust has evolved into a certain historical stage. The stratigraphy, lithologies, sedimentary environment, tectonics, slope surface development and geometrical features of a Danxia Landform all indicate its peculiar nature. Danxia Geomorphology and the related basic research will help to shed lights on Mesozoic and Post-Mesozoic crustal evolution, and to improve and enrich the geomorphologic theory. Danxia Geomorphology is initiated and developed in China. It is a subdiscipline of geomorphology and is approaching maturity step by step. It is also a contribution to science by the Chinese geoscientists.

Widely distributed in China, Danxia Landform has a great variety and is typical in development. The study of resources conservation, and environment cultivation, development and exploitation of Danxia Landform regions has a great practical significance in the development of the regional social economy. Especially as an excellent landscape, it is most worthy of development for tourism. The work for transferring the geomorphologic resources of Danxia Landform into economic, social and environmental benefits has increasingly attracted much attention of all walks of life. Therefore researchers working for the development of Danxia Landform for tourism are faced with a task of great significant in serving for production and construction.

(2) To strengthen the research on basic theory: In recent years, although many achievements have been scored in the basic research of Danxia Geomorphology, yet, in comparison with other branches of rock geomorphology especially with karst geomorphology, theories already established to explain Danxia Geomorphology are still very crude. In future, we should employ the more mature theories of petrology, structural geology, general geomorphology etc, and widely use modern research methods to explore into the conditions of formation, dynamic mechanism, and evolutionary process and stages of the Danxia Landform, in order to develop a sound theoretical system.

(3) To broaden the field of application study: Tourism development and tourism economic development have raised higher requirements in the application study of Danxia Geomorphology. In particular the scientific evaluation system, market analysis, tourism economic development and planning, and scientific popularization culture etc. are all new subjects the researchers must be faced with. It is predicted that the more practicable application study will play a vital role in future.

The results of application studies for tourism development as seen today are mostly still of a semi-finished nature, i.e., they are mostly concerned only with landscape description + proposal for development, and such studies are too flexible. The practicability as advocated by us is of three levels. The macroscopic level requires us to provide grounds for the government to made decisions of a strategic nature; the mesoscopic level requires us to work out the development layout and resources conservation plan; while the microscopic level requires that the microzone planning and design, project reviewing and construction can be put into practice. Therefore combination with other disciplines should be strengthened, including with applied geomorphology, tourist geography, tourist resources investigation and assessment, principles of tourism development, and tourism economics, so as to broaden the field of application study, to locate new growth points, to take part more directly, more intensely and more effectively in the tourism development practice, and to create more chances for the development of the discipline.

(4) To connect as soon as possible with international research: As far as we know to date, there are red beds on all of the old platforms in the world, and well-developed Danxia Landforms are present in the west part of the United States, Australia, Thailand and Brazil. However, as a geomorphologic type distinguished from other types, Danxia Landform has not yet been studied in foreign countries. Since many achievements have already been scored in the research of Danxia Geomorphology in China, the time has now come when we should exchange and cooperate with other countries in such research activities. Recently, through the various channels already established by the geological circle and the geographic circle, we should invite foreign researchers to take part in the cooperation and exchange activities, and to provide information concerning Danxia Landform and research progress in their countries. At an opportune moment we should organize and convene international symposium, so as to widen the influence of Danxia Geomorphology in the outside world, to connect our work with international research, and finally to establish the position of Danxia Geomorphology in international geomorphologic classification system.

参 考 文 献

1 冯景兰, 朱 翙声. 广东曲江仁化始兴南雄地质矿产. 两广地质调查所年报, 1928 (第1号)

2 陈国达. 广东之红色岩系. 国立中山大学理学季刊, 1935, 6 (4)

3 Kuota Chan. On the Subdivisions of the Red Beds of South-Eastern China. Bulletin of the Geological Society of China, 1938, 10 (8)

4 冯景兰. 关于"中国东南部红色岩层之划分"的意见. 地质论评, 1939, 6

5 陈国达、刘辉泗. 江西贡水流域地质. 江西地质汇刊, 1939 (第二号): 164 (图版3, 图C说明)

6 徐瑞麟. 广东北江地质之研究. 地质论评, 1937,4 (2)

7 曾昭璇. 仁化南部厚层红色砂岩区域地形之初步探讨. 国立中山大学地理集刊, 1943 (12)

8 吴尚时, 曾昭璇. 粤北之红层(The Red Beds in North Kwangtung). 岭南学报专号, 1946

9 吴尚时, 曾昭璇. 粤北红色岩系之地质与地形. 地学集刊, 1948, 6:13~45

10 周仁沾、杨超群. 关于东南沿海一带红色岩系划分的几点意见(节要). 地质论评, 1951,16 (1):131,132

11 曾昭璇, 李国珍. 华南砂岩地形特点. 地理知识, 1957 : 397~399

12 曾昭璇. 岩石地形学. 北京: 地质出版社, 1960. 45~57

13 曾昭璇, 黄少敏. 中国东南部红层地貌. 华南师范学报(自然科学版), 1978 (12)

14 李见贤(黄进). 广东省的地貌类型. 中山大学学报, 1961 (4): 72

15 曾昭璇, 黄少敏. 红层地貌与花岗岩地貌. 见: 中国自然地理·地貌. 北京: 科学出版社, 1980. 139~150

16 黄进. 丹霞地貌坡面发育的一种基本方式. 热带地理, 1982, 3 (2) : 107~134

17 张林源. 陇原上的"丹霞奇观". 甘肃日报, 1981-08-25(4)

18 刘振中. 武夷山形成与地貌发育特征. 南京大学学报(自然科学版), 1984 (3)

19 黄可光. 兰州安宁堡宫殿式类丹霞地貌浅析. 西北师范学院报(自然科学版), 1987 (3)

20 赵佩心. 承德丹霞地貌及旅游资源的开发利用. 地域研究与开发, 1988 (3)

21 陈致均, 黄可光. 甘肃丹霞地貌初探. 西北师范大学学报(自然科学版), 1989 (4)

22 陈传康. 风景地貌与地貌构景—地貌的旅游评价研究. 河南大学学报(自然科学版),1985 (1)

23 黄进主编. 第一届丹霞地貌旅游开发文集. 热带地貌, 1992(增刊)

24 黄进主编. 第二届丹霞地貌旅游开发学术讨论会论文集. 经济地理, 1994, 14(增刊)

25 黄进主编. 第三届丹霞地貌旅游开发学术讨论会论文集. 经济地理, 1996, 16(增刊)

26 彭华主编. 第四届丹霞地貌旅游开发学术讨论会论文集. 经济地理, 1998, 18(增刊)

27 彭华主编. 第五届丹霞地貌旅游开发学术讨论会论文集. 经济地理, 1999, 19(增刊)

28 彭华主编. 第六届丹霞地貌旅游开发学术讨论会论文集. 经济地理, 2000, 20(增刊)

29《地理学词典》编委会. 地理学词典. 上海: 上海辞书出版社, 1983

30 地质辞典办公室. 地质辞典(一): 普通地质构造地质分册(上册). 北京: 地质出版社, 1983

31 黄进. 丹霞地貌的旅游资源及其开发与保护. 见: 中国地理学会地貌与第四纪专业委员会编.地貌·环境·发展. 北京: 中国环境科学出版社, 1995

32 彭华主编. 广东人谈旅游文化. 广州: 广东旅游出版社, 1996

33 黄进. 中国丹霞地貌类型的初步研究. 热带地貌, 1991(增刊)

34 陈传康等. 丹霞风景名胜区旅游开发研究. 地理学报, 1990, 45 (3)

References

1 Feng Jinglan and Zhu Huisheng. Geology and mineral resources of Qujiang, Renhua, Shixing and Nanxiong, Guangdong Province. Annals of Guangdong and Guangxi Geological Survey, 1928 (No.1) (in Chinese)

2 Chen Guoda. Red rock series of Guangdong. Quarterly of National University of Sun Yat Sen (Natural Science Edition), 1935, 6(4) (in Chinese)

3 Kuota Chan. On the Subdivisions of the Red Beds of South-Eastern China. Bulletin of the Geological Society of China, 1938, 10 (8)

4 Feng Jinglan. Opinion in relation to "On the Subdivisions of the Red Beds of South-Eastern China". Geological Review, 1939 (6) (in Chinese)

5 Chen Guoda and Liu Huisi. Geology of Gongshui valley, Jiangxi. Geological Assembly of Jiangxi, 1939 (2) (in Chinese)

6 Xu Ruilin. A study on the geology of Beijiang, Guangdong. Geological Review, 1937,4(2) (in Chinese)

7 Zeng Zhaoxuan. Preliminary discussion on the landform in thickly bedded red sandstone region in southern Renhua. Geographic collection of National University of Sun Yat Sen, 1943(12):19~24 (in Chinese)

8 Wu Shangshi and Zeng Zhaoxuan. The Red Beds in north Kwangtung. Lingnan Journal, 1946 (special issue)

9 Wu Shangshi and Zeng Zhaoxuan. Geology and landform of red rock series in north Kwangtung. Geological Assembly, 1948 (6): 13~15 (in Chinese)

10 Zhou Renzhan and Yang Chaoqun. A few suggestions on the classification of red rock series on the southeastern coastal area (abstract). Geological Review, 1951, 16 (1):131, 132 (in Chinese)

11 Zeng Zhaoxuan and Li Guozhen. Characteristics of South China sandstone landform. Geographical Knowledge, 1957: 397~399 (in Chinese)

12 Zeng Zhaoxuan. Rock Geomorphology. Beijing: Geological Publishing House, 1960, 45~57 (in Chinese)

13 Zeng Zhaoxuan and Huang Shaomin. Red bed geomorphology in southern China. South China Teachers College Journal (Natural Science Edition), 1978 (12) (in Chinese)

14 Li Jianxian (Huang Jin). Geomorphologic Types of Guangdong Province. Journal of Zhongshan University, 1961, (4):72 (in Chinese)

15 Zeng Zhaoxuan and Huang Shaomin. Red bed geomorphology and granite geomorphology. In: Physical geography • Geomorphology of China.

Beijing: Science Press, 1980:139~150 (in Chinese)

16 Huang Jin. A basic form of slope development of Danxia Landform. Tropical Geography, 1982, 3(2):107~134 (in Chinese)

17 Zhang Linyuan. “Danxia Wonder” on Gansu Plateau. Gansu Daily, 1981-08-25(4) (in Chinese)

18 Liu Zhenzhong. The formation and geomorphologic development characteristics of the Wuyishan Mountains. Journal of Nanjing University (Natural Science Edition), 1984(3) (in Chinese)

19 Huang Keguang. A preliminary analysis on palatial type Danxia Landform in Anningbao, Lanzhou. Journal of Northwest Teachers College (Natural Science Edition), 1989(4) (in Chinese)

20 Zhao Peixin. Danxia Landform and the development and utilization of resources for tourism. Region Study and Development, 1988(3) (in Chinese)

21 Chen Zhijun and Huang Keguang. Preliminary investigation of Danxia Landform in Gansu. Journal of Norwest Teachers University (Natural Science Edition), 1989(4) (in Chinese)

22 Chen Chuankang. Landscape geomorphology and view construction with landform:tourism evaluation study of landform. Journal of He’nan University (Natural Science Edition), 1985(1) (in Chinese)

23 Huang Jin ed. Proceedings of the 1st symposium on Danxia Landform and tourism development. Tropical Geomorphology, 1992 (Supplementary Issue) (in Chinese)

24 Huang Jin ed. Proceedings of the 2nd symposium on Danxia Landform and tourism development. Economic Geography, 1994, 14 (Supplementary Issue) (in Chinese)

25 Huang Jin ed. Proceedings of the 3rd symposium on Danxia Landform and tourism development. Economic Geography, 1996, 16 (Supplementary Issue) (in Chinese)

26 Peng Hua ed. Proceedings of the 4th symposium on Danxia Landform and tourism development. Economic Geography, 1998, 18 (Supplementary Issue) (in Chinese)

27 Peng Hua ed. Proceedings of the 5th symposium on Danxia Landform and tourism development. Economic Geography, 1999, 19 (Supplementary Issue) (in Chinese)

28 Peng Hua ed. Proceedings of the 6th symposium on Danxia Landform and tourism development. Economic Geography, 2000, 20 (Supplementary Issue) (in Chinese)

29 Editorial Board of “Dictionary of Geography”. Dictionary of geography. Shanghai: Shanghai Dictionary Publishing House, 1983 (in Chinese)

30 Office of Dictionary of Geology, MGMR. Dictionary of geology (I): general geology and structural geology volume (Vol.1). Beijing: Geological Publishing House, 1983 (in Chinese)
31 Huang Jin. Resources of Danxia Landform for tourism and their development and protection. In: Commission on Geomorphologic and Quaternary Period, Geographical Society of China ed. Geomorphology • Environment. Development. Beijing: China Environmental Science Publishing House, 1995 (in Chinese)
32 Peng Hua ed. Cantonese speak about tourist culture. Guangzhou: Guangdong Tourist Publishing House, 1996 (in Chinese)
33 Huang Jin. Preliminary study on Danxia Landform types of China. Tropical Geomorphology, 1991(Supplementary Issue) (in Chinese)
34 Chen Chuankang *et al*. Study on tourism development of Danxia scenic resort. Acta Geographica Sinica, 1990,45(3) (in Chinese)

附 1 中国丹霞地貌分布及典型图片

Distribution of Danxia Landform in China and model Photographs

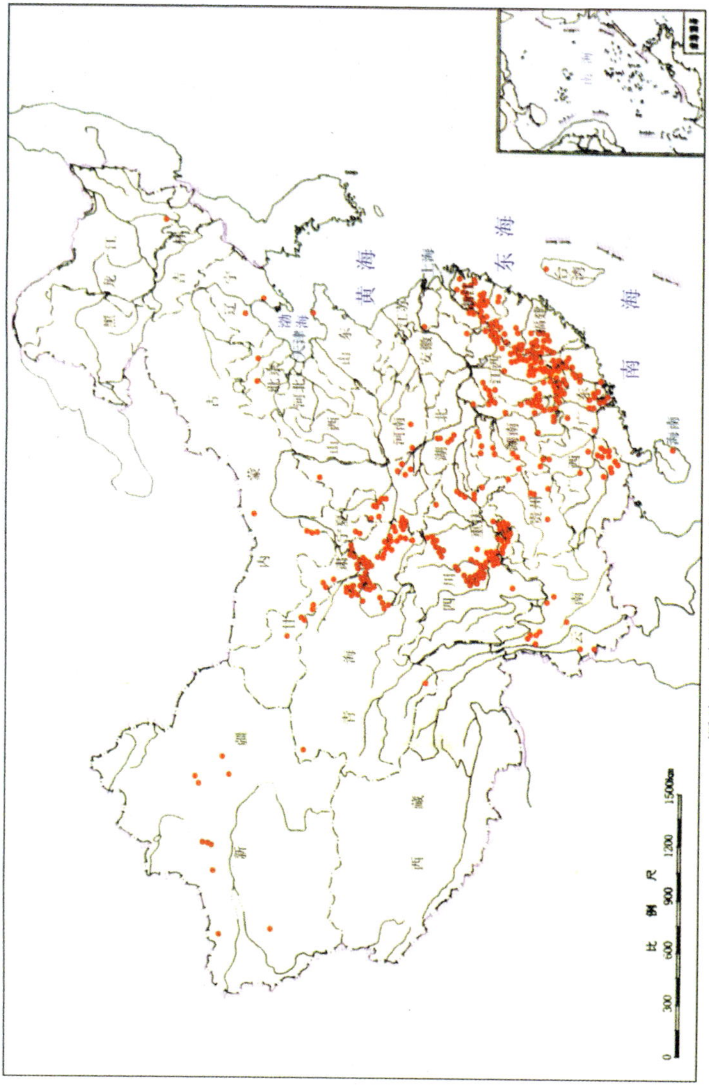

图片 1 中国丹霞地貌分布（根据黄进，1999）

Photograph 1 Distribution of Danxia Landform in China (modified after Huang Jin, 1999)

图片 2 丹霞山全景图(彭华绘制,1989)

Photograph 2 Overall view of Mt. Danxia (drawn by Peng Hua, 1989)

图片 3 丹霞山北部景观图（彭华绘制，1989）

Photograph 3 Landscape of the northern part of Mt. Danxia (drawn by Peng Hua, 1989)

图片 14　青海坎布拉半干旱区丹霞地貌群峰(彭华摄,1997)
Photograph 14　A group of peaks of the Danxia geomorphologic type in the Kanbula semiarid region, Qinghai (photographed by Peng Hua, 1997)

图片 15 江西龙虎山仙水岩一带老年期丹霞地貌群峰(摘自鹰潭画册)
Photograph 15　Senile Danxia geomorphology in the Xianshuiyan area, Longhushan, Jiangxi (from Yingtan Picture Album)

图片 16 贵州赤水十丈洞大瀑布(高 76.2m, 宽 80m)(引自赤水旅游图册)
Photograph 16 The Shizhangdong cataract (76.2m high and 80m wide) in Chishui, Guizhou (from Chishui Tourist Album)

图片 17 贵州赤水四洞沟瀑布群(右为第四瀑,下为第二瀑帘状瀑布)(彭华摄,1999)
Photograph 17 The Sidonggou group of waterfalls in Chishui, Guizhou (note the 4th waterfall on the right and the 2nd curtain-like water fall below)
(photographed by Peng Hua, 1999)

图片 18 福建泰宁水上一线天
Photograph 18 "a narrow strip of sky above water" in Taining, Fujian (by Peng Hua, 2000)

图片 19 福建泰宁上清溪峡谷
Photograph 19 The Shangqingxi canyon in Taining, Fujian (by Peng Hua, 2000)

图片 20 福建泰宁上清溪束窄峡谷
Photo 20 Shangqingxi restricted canyon in Taining, Fujian (by H. Peng, 2000)

图片 21 贵州习水板状河谷
Photo 21 Plate - like valley in Xishui, Guizhou (by H. Peng, 1999)

图片 22 广东丹霞山阳元石（天然石柱造型，高 28m，直径 7m）（刘家青摄，1993）

Photograph 22 A Stony Penis (28m high and 7m in diameter, formed by natural hoodoo column) in Mt. Danxia, Guangdong (photographed by Liu Jiaqing, 1993)

图片 23 阳元石所在的地貌部位（彭华摄，1993）

Photograph 23 Geomorphologic location of the Stone Penis (photographed by Peng Hua, 1993)

图片 24 广东丹霞山少阴石（天然石丘上的水蚀沟槽造型）（彭华摄，1999）

Photograph 24 A Stony Vulva (formed by rain channel on natural stone hill) in Mt. Danxia, Guangdong (photographed by Peng Hua, 1999)

3 黄进教授简介

图片 60 黄进教授近照
Photograph 60 Professor Huang Jin

黄进教授，广东省丰顺县人，1927年生，1952年中山大学地理系毕业，1984～1987年任中山大学地理系主任，1985～1991年任中国地理学会地貌与第四纪专业委员会副主任。1991年和北京大学陈传康教授发起并创立全国丹霞地貌旅游开发研究会，任第一和第二届理事长，组织召开了5届学术讨论会，主编出版了3本论文集，1998年退位，任名誉理事长。

黄进教授自20世纪60年代开始研究丹霞地貌，但系统研究开始于文化大革命以后的1978年，1982年发表了关于丹霞地貌坡面发育的经典论述。1989～1990年黄进教授得到中山大学自然科学基金的资助，1990～1994及1997～1999年又得到国家自然科学基金资助，先后对陕、甘、宁、川、黔、黑、冀、豫、鄂、湘、浙、皖、赣、闽、粤、桂、青、蒙、新、滇等省区进行丹霞地貌考察，到目前为止，在全国已经发现的500多处丹霞地貌中，黄进教授已实地考察了450余处。

黄进教授先后发表丹霞地貌研究论文40余篇，于60年代初第一次对丹霞地貌作了科学定义。80年代以来，他在系统考察的基础上，对丹霞地貌的坡面特征和发育规律、丹霞地貌特殊的风化作用、地壳上升与地貌发育过程的关系、减压节理与丹霞地貌发育的关系、丹霞洞穴地貌、丹霞喀斯特地貌、土林式丹霞地貌、低等生物对丹霞地貌坡面风化的影响、丹霞地貌的分布与区域差异、丹霞地貌分类等方面的研究均属于开创性工作，是当代中国丹霞地貌研究领域的学术带头人。

黄进教授至今出行仍然是一身中山装，一双解放鞋，一把黑雨伞，一个发白的帆布地质包，构成了他朴素鲜明的外表形象。而他一丝不苟的治学态度、坚韧不拔的献身精神、实事求是的工作作风、朴实正直的处世准则，确是激励后辈进取向上的宝贵财富。

图片 61 黄进教授在野外工作

Photograph 61 Professor Huang Jin working in the field

Brief Account of Prof. Huang Jin

Huang Jin was born in August 1927. His native place is Fengshun County of Guangdong Province. He graduated from Geographical Depart - ment of Zhongshan University in 1952, and was the Dean of Geographical Department in 1984 - 1987, and the Vice Chairman of Commission on Geomorphology and the Quaternary in Geographical Society of China. He, together with Prof. Chen Chuankang of Beijing University, initiated and set up The Danxia Landform and Tourism Development Research Society of China, in which he was president of the 1st and 2nd councils. He organized and convened five symposia, supervise the publication of three volumes of proceedings. He gave up the post in 1998 and was appointed as the honorary president of the council.

图片 62 70 岁高龄的黄进教授还可以攀崖

Photograph 62 Professor Huang Jin who is of advanced age of 70, can still climb up a cliff

volumes of proceedings. He gave up the post in 1998 and was appointed as the honorary president of the council.

Although Prof. Huang Jin began his study of Danxia geomorphology as early as in the 1960s, his systematic study did not start until 1978, i.e. posterior to the "Cultural Revolution". In 1982 he published his classic paper on the development of slope in Danxia geomorphology. Financially supported by 1989-1990 Natural Science Foundation of Zhongshan University and by 1990-1994 and 1997-1999 State Natural Science Foundation, Prof. Huang Jin carried out successively Danxia geomorphologic investigations in Shaanxi, Gansu, Ningxia, Sichuan, Guizhou, Heilongjiang, Hebei, He'nan, Hubei, Hu'nan, Zhejiang, Anhui, Jiangxi, Fujian, Guangdong, Guangxi, Qinghai, Neimenggu, Xinjiang and Yunnan. Up till now he has conducted on-the-spot examination in more than 450 out of the 540 Danxia geomorphologic sites found so far in China.

Prof. Huang Jin has earlier or later published over 40 papers concerning Danxia geomorphology. In the beginning of 1960s, he gave for the first time a scientific definition for Danxia geomorphology. Since the 1980s, through systematic investigation, he carried out research work on slope characteristics and development regularities in Danxia geomorphology, special weathering in Danxia geomorphology, relation between crustal uplift and the process of geomorphologic development, relation between decompressional joint and the development of Danxia geomorphology, Danxia cave geomorphology, Danxia karstic geomor-phology, earth forest type Danxia geomorphology, influence of lower organisms on slope weathering in Danxia geomorphology, distribution and regional difference of Danxia geomorphology, and classification of Danxia geomorphology, all of these are part of his initiative work. He is the academic foregoer in the Danxia geomorphologic research field in current China.

Up till now, when going out, Prof. Huang Jin is still in Chinese tunic suit and canvas shoes, carrying a black umbrella and a geologist's old canvas bag, displaying a simple and unadorned bearing. However, his meticulous scholarship, persistent devotion, practical and realistic style of work, and plain and honest ways of life, are really precious spiritual wealth encouraging the younger generation to keep forging ahead.